MAKING PIC® MICROCONTROLLER INSTRUMENTS AND CONTROLLERS

MAKING PIC® MICROCONTROLLER INSTRUMENTS AND CONTROLLERS

HARPRIT SINGH SANDHU

New York   Chicago   San Francisco   Lisbon   London   Madrid
Mexico City   Milan   New Delhi   San Juan   Seoul
Singapore   Sydney   Toronto

Library of Congress Cataloging-in-Publication Data
Sandhu, Harprit.
  Making PIC microcontroller instruments and controllers / Harprit Singh Sandhu.
      p.   cm.
  Includes bibliographical references and index.
  ISBN 978-0-07-160616-5 (alk. paper)
  1. Microcontrollers. 2. Electronic apparatus and appliances—Automatic control.
I. Title.
   TJ223.P76S37   2009
   621.39'16—dc22
                                                                                      2008049998

McGraw-Hill books are available at special quantity discounts to use as premiums and sales promotions, or for use in corporate training programs. To contact a special sales representative, please visit the Contact Us page at www.mhprofessional.com.

Making PIC® Microcontroller Instruments and Controllers

Copyright © 2009 by The McGraw-Hill Companies, Inc. All rights reserved. Printed in the United States of America. Except as permitted under the United States Copyright Act of 1976, no part of this publication may be reproduced or distributed in any form or by any means, or stored in a data base or retrieval system, without the prior written permission of the publisher.

1 2 3 4 5 6 7 8 9 0   FGR/FGR   0 1 4 3 2 1 0 9 8

ISBN 978-0-07-160616-5
MHID 0-07-160616-5

Sponsoring Editor
Judy Bass

Editorial Supervisor
David E. Fogarty

Project Manager
Vibha Bhatt,
International Typesetting and Composition

Copy Editor
Michael McGee

Proofreader
Priyanka Sinha,
International Typesetting and Composition

Indexer
Kevin Broccoli

Production Supervisor
Pamela A. Pelton

Composition
International Typesetting and Composition

Art Director, Cover
Jeff Weeks

Information contained in this work has been obtained by The McGraw-Hill Companies, Inc. ("McGraw-Hill") from sources believed to be reliable. However, neither McGraw-Hill nor its authors guarantee the accuracy or completeness of any information published herein, and neither McGraw-Hill nor its authors shall be responsible for any errors, omissions, or damages arising out of use of this information. This work is published with the understanding that McGraw-Hill and its authors are supplying information but are not attempting to render engineering or other professional services. If such services are required, the assistance of an appropriate professional should be sought.

About the Author

Harprit Singh Sandhu, BSME, MSCerE, is the founder of Rhino Robots Inc., a manufacturer of both robots and computer numeric-controlled machines.

CONTENTS

Preface xiii

PART I The PIC 16F877A   1

Chapter 1 An Introduction to the PIC 16F877A Microcontroller Unit   3
    The Microcontroller   7
    Special Precautions and Notes of Interest   9
    Data Sheets   9
    Some Useable PICS   11

Chapter 2 Getting Started: The Hardware and Software Setup   13
    The Programmers   14
    Loading the Software   15
    Using the Software in the Windows Environment   15
    Software Notes from MicroEngineering Labs   18

Chapter 3 Understanding Microchip Technology's PIC 16F877A: A Description of the MCU   21
    16F877A Microcontroller's Core Features   22
    Peripheral Features   23
    Configuring and Controlling the Properties of the Ports   29
    PORTA   29
    PORTB   30
    PORTC   31
    PORTD   32
    PORTE   33
    TIMERS   33

Chapter 4 The Software, the Compilers, and the Editor   35
    The Basic Compiler Instruction Set   35
    The PICBASIC PRO Compiler Instruction Set   37
    PICBASIC PRO Compiler   42

Chapter 5 Controlling the Output and Reading the Input   47
    General   47
    Programs That Create Output   48
    Programs That Read the Inputs and Then Provide Output   48
    Creating Outputs   48

Blink One LED *50*
Blink Eight LEDs in Sequence *51*
Dim and Brighten One LED *52*
LCD Display *53*
Controlling the Digital and Analog Settings *56*
Writing Binary, Hex, and Decimal Values to the LCD *56*
Reading a Potentiometer and Displaying the Results on the LED Bargraph *57*
A Simple Beep *60*
Advanced Exercise: Controlling an RC Servo from the Keyboard *63*
Reading the Inputs *67*
Read Keyboard and Display Key Number on the LCD *73*
Read One Potentiometer and Display Its 8-Bit Value on LCD in Binary, Hex, and Decimal Notation, Also Impress the Binary Value on the Bargraph *74*
Read All Three Potentiometers and Display Their Values on the LCD *76*
Adding the Kind of Flexibility That Defines Computer Interfaces and Demonstrates the Ability to Make Sophisticated Real-Time Adjustments *78*
Exercises *79*

Chapter 6 Timers and Counters   83

General *83*
Timers *84*
Timer0 *85*
Timer1: The Second Timer *93*
Timer2: The Third Timer *102*
Counters *104*
Exercises for Timers *112*
Exercises for Counters *112*

Chapter 7 Clocks, Memory, and Sockets   113

Sockets U3, U4, and U5: For Serial One-Wire Memory Devices *113*
Which EEPROM Type Should You Use? *115*
Socket U3—I2C SEEPROM *115*
Socket U4—SPI SEEPROM *117*
Socket U5—Microwire Devices *118*
Socket U6—Real-Time Clocks *120*
The LTC1298 12-Bit A-to-D Converter (*Also Used in Socket U6*) *124*
Sockets U7 (and U8) *126*

Chapter 8 Serial Communications: Sockets U9 and U10   131

When and How Will I Know if It Is Working? *134*
Using the RS485 Communications *137*

Chapter 9 Using Liquid Crystal Displays: An Extended Information Resource   139

General *139*
Using LCDs in Your Projects *142*
Understanding the Hardware and Software Interaction *143*
Talking to the LCD *144*

The Hardware   *144*
Setting Out Our Design Intent   *140*
Liquid-Crystal Display Exercises   *154*

PART II The Projects   **157**

Chapter 10 Using Sensors (Transducers)   **159**
General   *159*
The Most Basic Question We Must Answer Is...   *161*
Types of Sensors   *163*
Two Interesting Resources You Will Want to Investigate   *164*

Chapter 11 Conditioning the Input Signal   **165**
General   *165*
Alternating Current Outline   *166*
Direct Current Outline   *166*
Simple Switches and Other Contacts   *167*
Circuitry for Conditioning dc Signals   *169*

Chapter 12 Conditioning the Output Signal   **173**
General   *173*

Chapter 13 An Introduction to the Eight Projects   **177**
The Web Site   *177*
The Eight Techniques   *177*
Notes   *182*

Chapter 14 The Universal Instrument: A Background Discussion **183**
The Properties and Capabilities of a Universal Instrument   *183*
A Basic Temperature-Controlling Device   *184*
Notes   *186*

Chapter 15 Counting Pulses: A Programmable Tachometer   **187**
Project 1   *187*
Notes on Using Seven-Segment Displays   *199*

Chapter 16 Creating Accurate Intervals with Timers: The Metronomes **209**
Project 2   *209*
Timer0   *214*
Timer1   *224*
Timer2   *228*
The Timer2 Program   *229*
The Watchdog Timer   *230*

Chapter 17 Understanding the Counters: Counting Marbles   **233**
    Project 3   *233*
    Counting with an Escapement   *240*
    Some Real-World Notes   *243*
    Counting to a Register Using an Interrupt   *244*
    Counting Directly into an Internal Counter   *246*
    Using Timer1 in Counter Mode   *248*
    Special Notes for Timer1 Usage   *256*

Chapter 18 A Dual Thermometer Instrument   **259**
    Project 4   *259*

Chapter 19 An Artificial Horizon: A Table Surface That Stays Level   **269**
    Project 5   *269*
    Discussion   *270*
    Setting Up the Hardware Connections   *271*
    Building the Artificial Horizon Table   *275*
    Gravity Sensor Exercises   *277*

Chapter 20 Building a Simple Eight-Button Touch Panel   **279**
    Project 6   *279*

Chapter 21 Single Set Point Controller with Remote Inhibit Capability   **293**
    Project 7   *293*

Chapter 22 Logging Data from a Solar Collector   **301**
    Project 8   *301*
    Microcontroller Hardware   *304*
    Software   *306*

Chapter 23 Debugging   **315**
    General   *315*
    Debugging and Troubleshooting   *315*
    First Problem That Must Be Fixed: The Microcontroller Crystal Must Oscillate   *316*
    If the Chip Refuses to Run   *318*
    Using the PBP Compiler Commands to Help Debug a Program   *319*
    Commands That Can Provide Debug Output to a Serial Port   *319*
    Dumb Terminal Programs   *319*
    Solderless Breadboards   *320*
    Debugging at the Practical Level   *320*
    Configuring the 16F877A and Related Notes   *324*
    Settings   *326*
    Configuration   *326*
    Options   *327*
    Simple Checks   *327*

Some Programmer-Related Error Messages *328*
Things I Have Noticed but Have Not Figured Out (and Other Mysteries) *328*
Setting the Ports *329*

Chapter 24 Some Real-World Projects You Can Build   **331**

Conclusion   **335**

Appendixes   ***337***

Appendix A Setting Up a Compiler for One-Keystroke Operation   *339*

Appendix B Abbreviations Used in this Book and in the Datasheets   *341*

Appendix C Listings of PICBASIC PRO Programs on the Internet at melabs.com   *345*

Appendix D Notes on Designing a Simple Battery Monitor Instrument: Thinking about a Simple Problem Out Loud   *347*

Appendix E Using the Support Web Site to Help Make Instruments and Controllers   *349*

Index   *351*

PREFACE

The advent of the microprocessor in a small all-encompassing package and the availability of easy-to-use software have changed the scope of what can be easily accomplished in the engineering laboratory and on the hobbyist's workbench. The PIC® microcontrollers manufactured by Microchip Technology, Inc. of Tucson, Arizona form a formidable family of microcontrollers that can be used to fill myriad everyday needs. I have selected this family of microcontrollers for the projects in this book as a way of introducing the novice engineering student, serious hobbyist, and professional technician to the basic techniques that must be mastered to use these devices to make fairly sophisticated instruments and controllers.

The projects in this tutorial have been designed so they each emphasize the role of one specific technique for using these microcontrollers. Together, the eight projects give you the basic information and experience you require to design and build the unique instruments and controllers you will need for both your special and everyday needs.

The book is divided into two main areas of interest. The first part of the book introduces you to the PIC 16F877A in some detail so you know what is available in these microprocessors in the way of features. The second part of the book uses these features in the construction of the eight separate projects in detail. The 16F877A was chosen as the logic engine of choice in that this 40-pin IC has almost all the features one finds in the entire family of MCUs in the Microchip Technologies, Inc. offering. Once you understand the use of the 16F877A, you will be able to use the other microprocessors made by them and other manufacturers without difficulty. MicroEngineering Labs, Jameco, Solarbotics in Canada, and a large number of other vendors also provide a host of boards on which the 16F877A can be mounted.

The LAB-X1 board manufactured by microEngineering Labs was chosen as the basic board for all the experiments because it provides the user with a keyboard, an LCD display, a piezo speaker, and three potentiometers already mounted on the board and ready for use as part of the instruments we will make, and it can be used as a reliable test platform for the software we develop. It makes all the projects easier to assemble, experiment with, and modify as we develop them. The programs you create can be transferred to other microcontrollers in the family, with no modifications in some cases and minimal modifications in others. All you have to do is tell the compiler how you have changed the wiring, identify the microcontroller you are using, and it does the rest. Migration to MCUs made by other manufacturers should not present much difficulty either.

If you know very little about PIC microcontrollers and have never played with them, I strongly recommend you read my first book on microcontrollers, the *Tutorial and Resource Book for the LAB-X1*. This book is available from a number of sources, including me, microEngineering Labs in the U.S., Jameco, Solarbotics in Canada, and independent distributors all over the world. The first half of both books are similar in that they both concentrate on the properties of the microcontroller, but the first book will be much easier for you if you are a novice. This second book is targeted at readers who find my *Tutorial and Resource Book for the LAB-X1* too basic for their needs. Though this book covers the basic properties of the 16F877A, it does not linger on fundamental explanations and techniques. This text delves into advanced techniques and greater integration and so is more suitable for the advanced engineer, the university student, the technician, and the amateur engineer/hobbyist.

This book is not designed to make you an expert in machine language programming or to prepare you to become a technical expert on PIC microprocessors. It is designed to give you a solid understanding of what these logical engines can do for you in the laboratory/field and to show you how you can use them to do it. I will cover everything in a non-mathematical, even somewhat nontechnical format, so if you have a minimal understanding of things both mechanical and electronic, you can use these microcontrollers to do useful work. With this in mind, I have concentrated on using the PICBASIC PRO Compiler provided by the manufacturers of the board to generate the assembly-level machine code rather than spend time teaching you how to use assembly language. The compiler uses a dialect of BASIC that is easy to understand and very powerful in that it does everything you need done in the context of programming the PIC microprocessors. The code generated is both compact and fast. There are times when assembly language subroutines can be added to the programming to make an especially critical task execute faster, and if you need to do this, the compiler allows it and offers ample information on the subject. This workbook does not cover assembly language or the "C" language.

The second part of this book teaches you about building basic microcontroller-based instruments and controllers by showing you how to build eight separate devices. These eight projects are designed to help you deal with real-world situations and each one concentrates on one aspect of the techniques used to make controllers and instruments. Sensors of various kinds are utilized to see how they can be used with microcontrollers to get the results you want. A detailed discussion of the problems that will be encountered, and how they may be solved, are included.

The workbook starts out with the construction of a programmable tachometer that teaches you about the most fundamental techniques used in microprocessors: counting pulses. It also covers the use of seven-segment displays, so you will be comfortable using them since they are usually the best solution for minimal displays. Having mastered the techniques covered in this exercise, you will be able to interface to all sensors/sources that provide pulsed signals.

The ability to read real-world inputs and control real-world outputs is combined to first create simple interactions, and then more sophisticated ones. An entire chapter is devoted to reading inputs of all kinds. In this chapter, you will find the information needed to interface your MCU to the kinds of things you find in your home and office, as well as in the engineering laboratory. Both ac and dc signals are covered so you can bring the information into the MCU no matter what the source.

Another entire chapter is devoted to controlling outputs of all kinds. The chapter gives you the information you need to control the real world with an MCU. Both low-current TTL devices and high-current solid state switches and relays are covered.

By the time you get to the end of this book, you should be able to use these logic engines to create small yet fairly sophisticated systems that can do work that is both interesting and useful. From there, you will have the knowledge and confidence needed to take on more complicated and sophisticated tasks.

Since this book is designed to be a resource, you will see things repeated from time to time in more than one chapter. This will save you from having to read the entire book to get the information you need. By doing this, each chapter has been made as self-contained as possible (within reason). Segments of code in the programs also repeat themselves for the same reason.

A lot of what any intelligent machine does has to do with the software in the machine. The software in the machine is specific to that machine, but the knowledge that goes into the design of a good program is universal. In this book, we will look into how the software is designed for these small processors, as well as how to determine what must be done to make a machine function properly so it can more completely realize its potential for doing useful work.

PIC microcontrollers I selected the Microchip Technology family of PIC microprocessors as the focus of these notes for two reasons. First, Microchip Technology provides the most comprehensive line of microprocessors for the kinds of projects we are interested in, and second, the compiler for these processors provides almost the entire line of PICs with comprehensive support. All you have to do is tell the compiler which PIC you are using, and if the features you have been addressing in your project are available on that particular PIC, the compiler will do the rest. You will never have to buy another compiler if you stay with the very comprehensive Microchip Technology family of PICs.

The LAB-X1 The microEngineering Labs LAB-X1 board was picked as the support board because it provides a very comprehensive set of onboard features that will let us do almost all our software experimentation and development on it with ease. The PIC 16F877A was chosen because if has almost all the features you would expect to find on any small microcontroller. All the skills you will develop with this device will be transferable to any other system you decide to migrate to with ease. The board and the compiler together give you the most value for your investment.

Discounted package Arrangements have been made with microEngineering Labs to provide a comprehensive package of software and hardware items suitable for use with this book at a discount. A coupon to claim this discount is included at the end of this book.

HARPRIT SINGH SANDHU
Champaign, IL, U.S.A.

Internet support sites: www.encodergeek.com and
www.mhprofessional.com/sandhu

MAKING PIC® MICROCONTROLLER INSTRUMENTS AND CONTROLLERS

Part I

THE PIC 16F877A

AN INTRODUCTION TO THE PIC 16F877A MICROCONTROLLER UNIT

Making PIC-based instruments and controllers has to do with learning how to combine the various capabilities that single-chip microcontroller units provide and utilizing a coherent set of interactive functions that serve the purposes we have in mind.

A vast array of PIC microcontrollers is manufactured by the Microchip Technology Corporation of Tucson, Arizona. This tutorial is designed to introduce you to just one of these devices—and do so in a non-intimidating way, if you are technically inclined but not necessarily an electronic technician or electrical engineer. Getting to know this one device will expose you to most of the features provided in this entire family of devices and make you comfortable with selecting and using the device that most closely meets your needs.

The device I have selected is the PIC 16F877A microcontroller. This is a 40-pin device that contains most of the features you are likely to find in the family of microcontrollers that Microchip Technology manufactures. In general, the salient features we need to get comfortable with can be described as dealing with the following:

- Input capability
- Output capability
- Mathematical manipulations within 8/16-bit math
- Using timers
- Using counters
- Communications with digital and analog devices
- Dealing with displays
- Interacting with personal computers

The PIC 16F877A will let us get familiar with all the preceding items and more, and is an excellent choice for the first-time serious user. This particular MCU (microcontroller unit) is also very popular with robotic enthusiasts. A number of boards that incorporate this device are available in the general marketplace and on the Internet.

As novices, if we want to get familiar with the concept of building instruments and controllers based on PIC microprocessors, we need an easy-to-use yet sophisticated and versatile board to play with and test our nascent ideas on. Though, of course, it would be possible to design and build a board that would do that, we do not have the expertise or time at this juncture to do so. Therefore, I have selected the very popular LAB-X1 and the supporting PICBASIC PRO Compiler software as the basic platforms for the projects and ideas presented in this book. As you read this text, you will find that the system provides an easy-to-use and versatile platform for checking out your hardware and software ideas before committing to personal computer (PC) boards, wire, and solder.

MicroEngineering Labs, Inc., the manufacturers of the LAB-X1 board and the related PICBASIC PRO Compiler, maintain a very useful and helpful Web site that will be a tremendous aid to you as you learn about your LAB-X1 in particular, and the Microchip Technology Corporation PIC microcontrollers in general. Their Web site contains a large number of example programs, tutorials, and other technical information to help you get started using their boards. A large number of Internet-based Web sites also exist, which are dedicated to the use of PIC microcontrollers. You should bookmark those pertaining to your area of interest.

This resource supplements the information on the Internet both from the microEngineering Labs, Inc. site and from other sources. We will use the sample programs (modified for clarification as may be necessary) and other information on the Web in this book. This book provides extensive diagrams that will help you in designing your own devices based on what you will learn. The diagrams are on the support Web site as AutoCAD files and can be expanded on for your particular designs.

There are two basic aspects in familiarizing yourself with PIC microcontrollers—the hardware aspect and the software aspect. The LAB-X1 board is designed to provide you with the hardware platform you need to conduct your first software (and hardware) experiments with PIC microcontrollers. The PICBASIC PRO Compiler, provided by the manufacturers of the board to program the 16F877A and similar microprocessors, is both easy to use and powerful, and the code created is fast and efficient. Other compilers are available but not covered in this book.

If you have serious budget constraints, the software of choice for use with this board is the smaller Basic Compiler from microEngineering Labs. This compiler is available for about $100 (in 2008) and is not recommended by me for serious work. (A free copy of the PBP compiler is available on the microEngineering Labs Web site, which contains all the instructions in the full version of PBP but is limited to 30 lines of code. Even so, it can be used to effectively try out the powerful command structure of the language. The instructions for the language can be downloaded from the microEngineering Labs Web site at no charge. Before you make a decision in either direction, be sure to try out the free version of the compiler.)

On the other hand, if you have a serious interest in using PIC microcontrollers, the compiler I recommend is the PICBASIC PRO Compiler because it gives you the comprehensive power and ease of use you need to do useful, everyday, professional work rapidly. The PRO Compiler is available for about $250 (in 2008) and all the software discussed in this workbook was written with the PICBASIC PRO Compiler in mind. A comparative listing of the keywords provided with each compiler is provided in Chapter 4 on software and editors.

You will also need a hardware programmer that lets you transfer the programs you write on your PC to your PIC microcontroller. Programmers are available from microEngineering Labs for the parallel port, the RS232 serial port, and the USB port of your computer. These programmers make it a "one-button click proposition" to transfer your program from your computer to the microcontroller and run it without ever having to remove the MCU from the board. (The USB programmer is recommended by me.)

The editing software needed to write and edit the programs before transferring them to the programmer and onto the microcontroller is a part of the compiler package. Other editors are available at no charge from a number of other suppliers. None is a better choice than the editor provided.

The salient hardware features (with some repetition by categories listed) provided on the LAB-X1 are as follows:

The following *input* capabilities are provided:

- A 16-switch keypad
- A reset switch
- Three potentiometers
- IR (infrared detection capability) (no detector provided)
- Temperature sensing socket, (no IC provided)
- Real-time clock socket (no IC provided)
- Sockets for experimenting with three basic styles of "one wire" memory chips
- Serial interface for RS232 (IC is provided)
- Serial interface for RS485 (no IC provided)
- PC board holes are provided for other functions. See the microEngineering Labs, Inc. Web site.

The following *output* capabilities are provided:

- Ten-LED bargraphs with eight programmable LEDs
- 2-line × 20-character LCD display module
- A piezo speaker/horn
- DTMF capability (digital tones used by the phone company)
- PWM (pulse width modulation) for various experiments
- IR (infrared transmission capability) (no LED provided)
- Two hobby radio control servo connectors (no hobby servos provided)
- Sockets for experimenting with (also mentioned earlier)
    - Serial memories
    - A-to-D conversion with 12-bit resolution
    - Real-time clocks

The following *I/O interfaces* are provided:

- RS232 interface with IC
- RS485 interface, socket only (The chip is inexpensive and easy to obtain.)

You can investigate the use of the following three types of *Serial EEPROMs*:

- I2C
- SPI
- Microwire

The following *miscellaneous devices* are also provided:

- A reset button
- A 5-volt regulator
- A 40-pin ZIF socket for PIC micro MCU (the recommended PIC 16F877A IC is not provided)
- A jumper selectable oscillator from 4 MHz to 20 MHz
- An in-circuit programming/debug connector
- A prototyping area for additional circuitry
- A 16-switch keypad
- A socket for RS485 interface (device not included)
- A socket for I2C serial EEPROM (device not included)
- A socket for SPI serial EEPROM (device not included)
- A socket for Microwire serial EEPROM (device not included)
- A socket for real-time clock/serial analog to digital converter (devices not included)
- A socket for Dallas 1620/1820 time and temperature ICs (devices not included)
- An EPIC in-circuit programming connector for serial, USB, or parallel programmer

All in all a very comprehensive, well thought out, and useful experimental platform suitable for our investigations.

The board is available assembled, as a kit, or as a bare PCB. The board is 5.5 inch × 5.6 inch.

Not all the features I have mentioned here are completely implemented, but, as I stated, sockets or PC board pin holes are provided for all of them. You may not have to make any soldering additions to the board to use the features you are interested in, but you do have to purchase the additional IC chips if you are interested in their use. The standard version of the board as shipped to you includes the following

- The assembled board
- Software diskette
    - PDF schematic of LAB-X1
    - Sample programs
    - Editor software
    - Additional support information (on the Web site)
    - (The 40-pin PIC microcontroller is not included.)
    - As received, the board is configured to run at 4 MHz.

The Microcontroller

The PIC 16F877A microcontroller (which is a necessary component on the board) is not provided, because each of the compatible PIC microprocessors available have varying features and you may want to select a unit that suits the application you have in mind. We will be using the recommended PIC 16F877A microcontroller for all our experiments. If you want to use another processor, be sure to check it for pin-to-pin compatibility with the LAB-X1 board on the Web. Datasheets can be downloaded for all the microcontrollers at no charge from the Internet. A large number of PICs can be used in the LAB-X1. See the list at the end of this chapter.

THE SOFTWARE COMPILER

The PICBASIC PRO BASIC software compiler (to be purchased separately), produced by microEngineering Labs, offers the functions needed to control all aspects of the hardware provided by Microchip Technology as a part of their large PIC offering. All the functions available on the PIC 16F877A microcontroller we will be using are accessible from the software. The PICBASIC software will create programs for almost the entire family of PIC microcontrollers. You will be able to use this compiler for all your future projects if you stay with the Microchip MCUs. (It is, all in all, a very worthwhile investment.)

ADDITIONAL HARDWARE

The following hardware can be added without making any modifications to the board. These hardware items fit into sockets or onto pins that are provided on the LAB-X1 as shipped. Not all devices can be mounted simultaneously in that some addresses are shared by the sockets provided. In our experiments, we will populate only one empty socket at a time in order to make sure no conflicts arise. (There is no need to use more than one device simultaneously for any one experiment, so this will not be a problem.)

- Memory chips
  - I2C memory chip
  - SPI memory chip
  - Microwire memory chip
- 12-bit A-to-D converter chip
  - NJU6355
- Real-time clock chips
  - DS1202
  - DS1302
  - LTC1298
- Thermometer chip
  - DS1802

- Serial interface chip
  - RS485
- RC servos (Two hobby R/C servos can be controlled simultaneously.)
- The LAB-X1 provides two sets of pins for the R/C servos. All standard model aircraft servos can be employed and you can use either one or two servos. (Using these is essentially an exercise in creating pulse width modulated signals and profiles that meet the standards used in the radio control hobby industry.)

40-PIN DEVICES

All 40-pin MCUs offered by Microchip can be accommodated in the 40-pin ZIF socket provided on the board. (Check for compatibility with the pin layout before selecting/buying your MCU. See Figure 1.1. The recommended PIC 16F877A we are using is an excellent choice for learning.)

BREADBOARDING AND EXPANSION

All of the MCUs 40 pins (on the LAB-X1) have been provided with extra PC board holes that can be used to extend the signals from these pins to an off-board location for further experimentation. The extensions are easily made with standard 0.1-inch on-center pins with matching cables with headers.

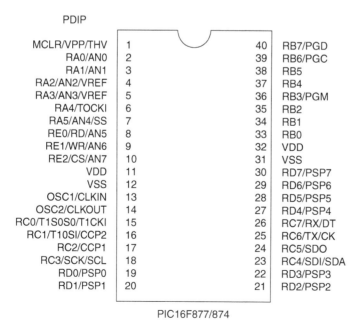

Figure 1.1 Pin out designations for the 40-pin 16F877A PIC microcontroller.

A small breadboard space is provided on the LAB-X1 itself to allow the addition of a limited number of hardware items you may need to experiment with.

See the support Web site regarding the availability of readymade headers and cables and so on, for use with the LAB-X1. (These are the devices I needed and made up to allow me to experiment with the 16F877A when the circuitry I required was not available on the LAB-X1.)

Special Precautions and Notes of Interest

These caveats could have been placed later in this book but are included here to encourage you to think about the programmer best suited to your needs.

Pin B7 on the 16F877A is connected to a programming pin on the EPIC parallel programmer at all times, and the programmer forces this pin high. If you are using this pin in your experiment and you decide it must be *low*, you must disconnect the EPIC programmer to release this pin. If you are using a serial or USB programmer, it can be left connected to the LAB-X1 at all times. The major benefit of using the USB or parallel programmer is that it frees up your computer's serial port for communications to the LAB-X1.

Resistor R17, which is connected to the keypad, is of no consequence to the operation of the LAB-X1. It is needed for some (PIC) programming functions and can be ignored (for our purposes).

Data Sheets

The hardest part of learning how to use these microcontrollers is understanding the huge datasheets. Since each datasheet is similar but different from every other datasheet, you are advised to select one or two microcontrollers to get familiar with and then use them for all your initial projects. In this workbook, the two discussed/mentioned are the PIC 16F84A (this chip will not fit in the 40-pin socket provided but is a good alternate choice for the cost-conscious), for your small projects, and the PIC 16F877A, for larger more comprehensive projects. Each of these uses flash memory and can therefore be programmed over and over again with your programmer and a programming socket. The processor you select will be determined by the kind of I/O and internal features you need, and the availability of inexpensive OTP (one time programmable) equivalents if you plan to go into production.

A lot of the information in the datasheets is more complicated and detailed than we need to worry about at this time, and we can do a lot of useful work without understanding it in every detail. Our main interest is in what the various registers are used for and how to use them properly and effectively. The timing diagrams and other data about the internal workings of the chips are beyond what we need to understand at the level of this resource book. Our interest is in being able to set the various registers in the system so we can activate the features we need for each particular project. Understanding timers and counters is a part of this. The entire interaction of the microcontroller with its environment is determined by the I/O pins and how they are configured, so knowing how to configure the I/O is very important.

The datasheets are available as PDF (page description format) files on the Internet from the microEngineering Labs Web site or from the Microchip Technology Web site. These should be downloaded onto your computer for immediate access when you need them. Keeping a window open specifically for this data is very handy. Even so, you will want to print out some of the information to have it in "in your hands."

Of particular interest are the areas of the datasheets that will support our needs as they apply to the following areas:

- Understanding and becoming familiar with what has already been defined by the Compiler software as it relates to the software
- Getting familiar with the addressing and naming conventions used in the datasheet
- Understanding the use of the various areas of memory on the MCU
- Learning how to assign and use the I/O pins to our best advantage
- Understanding how to use the PICBASIC PRO software effectively
- Getting familiar with the general register usage as it is implemented for the control of the timers and counters

A FAST INTERNET CONNECTION IS PRETTY MUCH A MUST

You absolutely have to have an Internet connection, and it is very helpful to have more than a standard phone line connection, so get the fastest connection you can afford. You need the connection because so much of the information you need is on the Internet. A cable modem is strongly recommended. If you and a couple of neighbors can get together and form a local area network (LAN) and share a wireless modem set up, it becomes a really inexpensive way to get fast Internet service. The Wi-Fi signals have no problem reaching all the apartments in a small building and sometimes even the house next door. Relatively inexpensive amplifiers and repeaters are available to increase signal strength where necessary.

DOWNLOADING DATA SHEETS

One of the first things you need to download is the datasheets for the PIC16F87XA. You will, in all probability, end up using the smaller and less expensive PIC 16F84A for a lot of your initial projects, so it might be best to download the information for that microcontroller while you are at it. As mentioned before, these files are available from the Microchip Technology Web site and the information is free. However, the two documents consist of about 400 pages, so you probably will not want to print it all out. You will, however, want to have some of the more commonly used information printed so you can refer to it whenever necessary. The rest should be stored on your computer so you can call up or search for what you need when necessary.

The Microchip Technology Corporation Web site is www.microchip.com.

Click "Support" on their Web site to find what you need. The items are easy to download. Just follow the instructions provided on the site.

Note *Much of the information you will need is provided on the Web site that supports this book. But even so, it can't replace a fast Internet connection.*

Some Useable PICS

The following 40-pin PICs will work in the LAB-X1 (as of June 2008). Others may work as well. Check these ICs for the features you need for your particular application, and then select the one that provides the best match.

PIC16C64(A), 16C65(B), 16C662, 16C67, 16C74(AB), 16C765, 16C77, 16C774, 16F74, 16F747, 16F77, 16F777, 16F871, 16F874, 16F874A, 16F877, **16F877A,** 16F914, 16F917, 18C442, 18C452, 18F422, 18F4320, 18F4331, 18F4410, 18F442, 18F4420, **18F4431**, 18F4439, 18F4455, 18F448, 18F4480, 18F4510, 18F4515, 18F452, 18F4520, 18F4525, 18F4539, 18F4550, 18F458, 18F4580, 18F4585, 18F4610, 18F4620, 18F4680

We are considering the 16F877A.

The 18F4331 family optimizes motion control for encoded motors.

Features provided on each MCU by the manufacturer vary from chip to chip. A chart of comparative features is maintained by microEngineering Labs on their Web site.

2

GETTING STARTED: THE HARDWARE AND SOFTWARE SETUP

This chapter deals with the minimum hardware and software you need to get started, and what you must do to set it up and get it ready for use.

The following is a list of the hardware needed, and what comes with each item:

- The LAB-X1 board
  - Software diskette
- Power supply for LAB-X1
  - A wall-mounted transformer (9 to 16 VDC at 1 amp)
- USB port programmer for the board (or other programmer)
  - Software diskette
  - A 10-pin parallel cable with 2 × 5 connectors (programmer to LAB-X1)
  - USB cable from computer to programmer
- Power supply for the programmer (not needed for a USB programmer)
  - Wall-mounted transformer
- A PIC 16F877A microcontroller or board-compatible 40-pin MCU

The following is a list of the software required:

- PICBASIC PRO Compiler
- MicroCode Studio editor software for writing the programs
- Programmer software (comes with the programmer)

The information needed is as follows:

- Datasheet for PIC 16F877A microcontroller (downloaded from the Internet)

You should already have the following computer equipment:

- Wintel computer (IBM-PC or compatible)
  - Hard drive
  - Printer
  - A Windows operating system
  - Access to the Internet. A broadband connection is recommended.

The Programmers

Three programmers are offered by microEngineering Labs. One uses the parallel port, one uses the USB port, and the third uses the serial port. The operation of the three programmers is almost identical (as far as the user interface is concerned). We will use a USB port programmer for all our experiments because it is more convenient to use than the others since it does not need a power supply. An important added bonus is that if frees up the COM port for use with the computer (the parallel programmer allows this also).

NOTES ON USING PROGRAMMERS

The USB programmer does not need a power supply or wall transformer. It gets its power from the USB port. Using a USB port frees up the serial port for your experimentation, which is important because most new computers have only one serial port. The PC serial port connects to the LAB-X1 serial port for communication experiments.

For the serial port and parallel port programmers, first plug the 16-volt power cord connector into the programmer and then into the wall socket. The USB programmer also needs to be connected but does not need a power supply connection. If you do not have power to the programmer when you start the programming software, the software will not be able to see the programmer and an error message will be displayed. The software will report that it could not find the programmer.

It is best to start the programmer software from the MicroCode Studio editor window. If you do it this way, the programmer being used is selected automatically and the program you are working on in the MicroCode editor window is transferred automatically to the compiler software and onto the MCU on the LAB-X1 board. It can be set up to be a "one mouse click" operation. See this book's Appendix A for more information.

Insert the microcontroller into the programming socket immediately before you begin programming the microcontroller (if you are programming an MCU that is not on the LAB-X1). This applies only if you are programming a loose microcontroller. If you are programming a microcontroller plugged into the LAB-X1, it can be left in the board all the time. (The only exception is for the parallel port programmer, because the B7 pin is pulled low by this programmer and thus will interfere with your program. If you plan to use this pin, you must unplug the programmer between programming sessions. This is mentioned again as a reminder and caution.)

The sequence to create a program inside a microcontroller is as follows:

1. Write the program in the MicroCode Studio editor environment.
2. Compile the program.
3. Program the device.
4. Use the device.

Steps 2 to 4 can be combined into one keystroke. See Appendix A.

Loading the Software

The following pieces of software will be provided with the various components you acquire:

- *PICBASIC PRO Compiler* software and manual
- *USB Port Programmer* software and manual (or software for whatever programmer you decide to buy)
- *MicroCode Studio* (the editor) on disk or downloaded from the Internet

The DOS environment is archaic and can be difficult for users not familiar with it. You do not have to deal with DOS to use the hardware and software we will use. Everything can be done from the Windows environment. Therefore, *DOS will not be discussed*. If you need to use DOS, a section at the beginning of the PICBASIC PRO Compiler manual will tell you what you need to do.

The first manual you need to understand is the manual for the PICBASIC PRO Compiler. This manual covers use of the software in the DOS environment. I suggest you ignore the first pages of the manual and instead read the paragraph (below) that I have written on how to run everything under the Windows environment. Once you are familiar with how the system works, you can go back and learn how to use the software in the DOS environment. The DOS environment provides a number of things that may be useful, and you will want to know about these as you become more and more proficient with microcontrollers.

Using the Software in the Windows Environment

The first question that needs to be answered in almost every endeavor is always "What do I need, what do I have to do, and what will it cost to get the job done?" Accordingly, we will address this now.

Let's assume you already have an IBM-PC with a suitable Windows operating system and that you know how to use it. Your computer needs the following capabilities to let you access the hardware and software you are going to use it with. In this book, I will deal *exclusively with the IBM-PC in a Windows environment.* The software is not available for the Macintosh. The following outlines what you will need:

- A 3.5-inch floppy drive (if the software is provided on 1.4-MB 3.5-inch floppies only and you must read if off a diskette for the system to work right). (Incidentally, you cannot copy the software to a CD-ROM and work from there. It will not work for some software.) If the software is provided on CD-ROM, you need a CD-ROM drive.
- A hard disk with about 5 MB of free space. This is for software storage and general workspace.
- A serial port (COM1 or COM2) if you will be using the new serial programmer, and a USB port if you will be using the USB programmer.
- LAB-X1 Experimenters board $195
- 16F877A Microcontroller (not part of LAB-X1) $10
- USB programmer $99
- PICBASIC PRO Compiler $250
- Miscellaneous electronic items for experimentation $100 allowance

At the end of this book there is a discount voucher from microEngineering Labs worth about $70 off their *Developers Bundle*. A description of what you get is included with the voucher. You have to use the actual voucher in the book; a photocopy will not be accepted.

The $100 allowance is for the sensors, servos, and other miscellaneous electronic components. It covers the need to purchase memory- and time-based components that are socketed for, but are not provided as a part of, the needs mentioned previously. You may decide that you do not need to experiment with these at this time. The allowance provides for almost everything you need to experiment with the LAB-X1 board.

Also provided by microEngineering Labs, Inc. are a number of other preassembled boards for experimentation and educational purposes. They are listed next and may be more closely suited to the instrument you are interested in creating. All can be used.

- The LAB-*X1 Experimenter's board* we are considering.
- The LAB-X2 Experimenter's board for custom circuits.
- The LAB-X20 Experimenter's board for 20-pin devices.
- The LAB-3 Experimenter's board for 18-pin devices.
- The LAB-4 Experimenter's board for 8- and 14-pin devices.
- The LAB-XT Experimenter's board for telephone technology–related investigations.
- The LAB-XUSB Experimenter's board for building USB interfaces and peripherals.

In this book, we will consider the LAB-X1 only. This board provides a 2-line-by-20-character display, which is very useful in the learning environment in that it can allow you to see what is going on in the system as you experiment (if you program your programs to display the appropriate information).

Start out by opening a new folder on your desktop and labeling it **PIC Tools**. We will *store everything that has to do with all our projects in this one folder.* We are opening this folder on the desktop now but you can move it to wherever you like in the future. Doing it this way avoids having to make a decision as to where to locate the folder, and the folder is right in front of you when you start your computer and the desktop appears.

Open the PIC Tools folder, and in it open folders—one for each of the items/applications—that we are going to be working with in this folder. Name these folders as follows:

- MicroCode Studio
- USB Programmer (or whichever unit you decide on)
- PICBASIC PRO Compiler
- LAB-X1

Put the MicroCode Studio diskette/CD in the disk drive and open it.
Copy all files to the MicroCode Studio folder.
Eject the diskette/CD and store it in a safe place.
Put the programmer diskette/CD in the disk drive and repeat the steps listed previously for the software in this package. Repeat the process for all the diskette/CDs.

Put a shortcut for the MicroCode Studio program on your desktop. This is the only shortcut you need when you want to create programs for your MCUs. All other functions of the system can be accessed from the window of this editor.

It's a good idea at this time to go to the microEngineering Web site and download the information on the LAB-X1 experimenter board, storing it in your LAB-X1 Tools folder. Useful example programs abound in these files and cutting and pasting from these to the programs you are writing will save you a lot of time. These programs are also on the book's support Web site.

If you are familiar with and have *information for the Basic Stamp*, it would be a good idea to add these files to this file so all your microcontroller information is in one place.

As a general rule, you will never see the compiler as such. It is called from the MicroCode Studio editor screen, does its work on compiling the program it is asked to compile, and then disappears into the background ready for the next compilation request. The errors displayed after a compilation are generated by the compiler. If all goes well, there should be no errors and you should get a message telling you the compilation was successfully performed. The new HEX file just generated will appear in the directory listing the next time you open a file. The HEX file will have the same name as the text file it was compiled from. The PICBASIC PRO Compiler manual covers how all this is done in more detail:

```
Source file    Untitled.bas
HEX file       Untitled.HEX
```

It should be noted that the HEX file is not created until all the syntax errors that the compile can find have been *eliminated by you*. After a successful compilation of the code, there may still be errors in the programming itself that will need to be addressed as you debug your work.

The address of the microEngineering Web site is www.melabs.com/index.htm.

If you have a CD-ROM burner on your computer (and if you do not, you should get one), it is well worth your time to now *copy the entire unadulterated PIC Tools folder to a CD-ROM* for safekeeping. Data on a CD-ROM is much more secure than the data on a floppy disk or diskette and the best time to make a copy of it is right now before you make any changes to any of the data you received from the vendors.

For the purposes of general discussion and experimentation, we will always call the example *program that is being manipulated "Untitled",* while the text file that is the body of the program will be called

```
Untitled.bas
```

This is the file the compiler compiles for the microcontroller you are using to create the HEX file it needs.

The HEX file created from this program by the compiler will be referred to as

```
Untitled.HEX
```

We do it this way because *every time you compile and run a program, the system automatically saves the program* to disk at the same time. This means that you lose the old program and cannot go back to it. If you are working with a complicated program, this can become a real problem because there are lots of good reasons to go back to the way things were. So, to avoid this pitfall, every time you load a program from disk, *first save it as* Untitled.bas and then play with it all you want. Then, when you have a viable program, save it to the name that is appropriate for it. Load the next program and change its name to Untitled.bas and so on. I would even go so far as to recommend that you save each version of your program with a version designation so that Blink.bas will be worked on as "Untitled.bas," and then re-saved to disk as "BlinkV1.0.bas," while "BlinkV1.0.bas" would be worked on as "Untitled.bas" and then re-saved as "BlinkV1.1.bas," and so on. Though some tedium is involved in doing this, I can assure you it will save you many headaches in the long run.

Note *The HEX files created by the PBP compilers can be loaded into the PIC microcontrollers with other software/loaders. It is not imperative that hardware programmers be used.*

Software Notes from MicroEngineering Labs

The following are some notes from microEngineering Labs regarding the software provided by them:

- Locate the CDs labeled *PICBASIC PRO Compiler* and *microEngineering Labs Programmer.*

- The compiler upgrade CD is identical to our full-version CD (with the exception of the red stamp). You can run the setup program to begin the installation process. It will install PBP, and also offers to install both MPLAB and MicroCode Studio.
- The programmer CD is the same for all the programmers. The software automatically finds whichever programmer you have connected.
- The LAB-X1 materials can be found on the PBP CD (\labx\LABX1) or on the LAB-X1 product page on our support Web site. A disk is no longer shipped by microEngineering Labs with the X1.
- The compiler ships on CD with a printed manual. Besides the compiler, the CD includes MPLAB (from Microchip), MicroCode Studio (from Mecanique), LAB-X sample programs and documentation, and selected PICmicro datasheets.
- All the programmers include software on CD. The meProg software is compatible with any microEngineering Labs programmer hardware.
- The LAB-X boards include a schematic and brief description printed on paper as well as sample programs, applicable datasheets, and details on their use.
- The latest compiler (2.50) is shipped with an enhanced version that offers 32-bit long variables and signed 32-bit math operations. This is compatible with PIC18 series microcontrollers only.

3

UNDERSTANDING MICROCHIP TECHNOLOGY'S PIC 16F877A: A DESCRIPTION OF THE MCU

PIC microcontrollers are manufactured by the Microchip Technology Corporation of Chandler, Arizona. The following address and Web site may be used to contact Microchip Technology. The Web site provides downloads for the datasheets.

Microchip Technology Corporation Inc.
2355 West Chandler Blvd.
Chandler, Arizona, USA 85224-6199
(480) 792-7200
(480) 899-9210 FAX
Web site: www.microchip.com

MicroEngineering Labs, Inc. maintains a very useful and helpful Web site that will also be a tremendous aid to you as you learn about the PIC microcontrollers by using their LAB-X1.

MicroEngineering Labs
Box 60039
Colorado Springs, CO 80960-0039
(719) 520-5323
(719) 520-1867
e-mail: Support@melabs.com
Web site: www.melabs.com/index.htm

MicroEngineering Labs
1750 Brantfeather Grove
Colorado Springs, CO 80960

We will be using *the 16F877A Microcontroller* in the LAB-X1 board. Not all the features described in the 220-page datasheet of the 16F877A will be addressed in the exercises to follow, but enough will be discussed to give you the confidence and understanding you need to proceed on your own. In more technical terms, this MCU has the following features

16F877A Microcontroller's Core Features

(Reduced from descriptions provided by Microchip Technology, Inc.)

High-performance RISC CPU

Operating speed: DC - 20 MHz clock input

DC - 200 ns instruction cycle

Up to 8K × 14 words of FLASH program memory

Up to 368 × 8 bytes of data memory (RAM)

Up to 256 × 8 bytes of EEPROM data memory

Interrupt capability

Power-on reset (POR)

Power-up timer (PWRT) and oscillator start-up timer (OST)

Watchdog timer (WDT) with its own on-chip RC

Programmable code-protection

Power saving SLEEP mode

Selectable oscillator options

Low-power high-speed CMOS FLASH/EEPROM technology

Fully static design

In-Circuit Serial Programming (ICSP) via two pins

Single 5-volts In-Circuit Serial Programming capability

In-Circuit Debugging via two pins

Processor read/write access to program memory

Wide operating voltage range: 2.0 volts to 5.5 volts

High sink/source current: 25 milliamp

Commercial and industrial temperature ranges

Low-power consumption

Peripheral Features

The following are peripheral features of the 16F877A Microcontroller:

- Timer0: 8-bit timer/counter with 8-bit prescalar.
- Timer1: 16-bit timer/counter with prescalar (It can be incremented during sleep via an external crystal/clock.)
- Timer2: 8-bit timer/counter with an 8-bit period register, prescalar, and postscalar
- Two PWM modules (maximum resolution is 10 bits)
- 10-bit multichannel analog-to-digital converter
- Synchronous serial port (SSP)
- Universal synchronous asynchronous receiver transmitter (USART)
- Parallel slave port (PSP) 8 bits wide
- Brown-out detection circuitry for Brown-out reset (BOR)

This MCU is described in profuse detail in a 220-page datasheet you can download from the Microchip Web site at no charge. The datasheet is a PDF document that you should have available to you at all times (maybe, even open, in its own window, ready for immediate access), whenever you are programming the 16F877A. The software you need in order to read (but not write) PDF files is also available at no charge on the Web. You should have a copy of the latest version (9) of this very useful software (Adobe Reader) on your computer.

We will not cover the entire 220-page datasheet in these exercises, but the most commonly used features of the MCU (especially the ones relevant to the LAB-X1 and those needed for our instruments and controllers) will be discussed. After doing the exercises, you should be comfortable with reading the datasheet and finding the information you need to get your work done.

In our particular case, on the LAB-X1 board, the MCU is already connected to the items on the board. Therefore, if you want to use the LAB-X1 for your own hardware experiments, you must use the MCU pins in a way that is compatible with the components that are already connected to them. Often times, even though a pin is being used in the LAB-X1 circuitry, you can drive something else with it without adversely affecting your experiment (depending on the load being added). The following is a list you can refer to in order to quickly determine if the pin and port you want to use is free, or discover how it is being used.

PORTA	PIN#	USAGE		PORTA HAS ONLY SIX EXTERNAL PINS
PORTA.0	2	5K ohm Potentiometer	0	Memory chips
PORTA.1	3	5K ohm Potentiometer	1	Memory chips
PORTA.2	4	Used by clock chips		
PORTA.3	5	5K ohm Potentiometer	2	Used by clock chips, memory

PORTA.4	6	This specific pin has special pull-up needs!	No analog function
PORTA.5	7	Free for A-to-D conversion	Memory chips
PORTB			
PORTB.0	33	Keypad inputs	
PORTB.1	34	Keypad inputs	
PORTB.2	35	Keypad inputs	
PORTB.3	36	Keypad inputs	Progr'g device
PORTB.4	37	Keypad inputs	
PORTB.5	38	Keypad inputs	
PORTB.6	39	Keypad inputs	Progr'g device
PORTB.7	40	Keypad inputs	Progr'g device
PORTC			
PORTC.0	15	Servo/Clock	
PORTC.1	16	Clock chips	Memory chips, servo/clock, HPWM
PORTC.2	17	Piezo speaker	HPWM
PORTC.3	18	Clock chips	Memory chips, servo/clock
PORTC.4	23	Used with Memory chips	
PORTC.5	24	Clock chips	A/D conversion, memory chips
PORTC.6	25	Transmit serial communications	RS232C
PORTC.7	26	Receive serial communications	RS232C
PORTD			
PORTD.0	19	LCD and LED bargraph	
PORTD.1	20	LCD and LED bargraph	
PORTD.2	21	LCD and LED bargraph	
PORTD.3	22	LCD and LED bargraph	
PORTD.4	27	LCD and LED bargraph	
PORTD.5	28	LCD and LED bargraph	
PORTD.6	29	LCD and LED bargraph	
PORTD.7	30	LCD and LED bargraph	
PORTE	**PORTE Has Only 3 External Pins**		
PORTE.0	8	LCD writing controls	
PORTE.1	9	LCD writing controls	
PORTE.2	10	LCD writing controls	Communications

Other Pins

Pin 1	MCLR	Microprocessor reset pin pull-up	Programming
Pin 11	Vdd	Logic power 5VDC	Has no other use
Pin 12	Vss	Logic ground	Has no other use
Pin 13	OSC1	Oscillator	Has no other use
Pin 14	OSC2	Oscillator	Has no other use
Pin 31	Vss	Logic ground	Has no other use
Pin 32	Vdd	Logic power, 5VDC	Has no other use

Re-listed in serial order, the pins are used as follows:

Pin 1	MCLR	Processor reset pin, pull up	Progr'g device
Pin 2	PORTA.0	5K ohm Potentiometer 0	
Pin 3	PORTA.1	5K ohm Potentiometer 1	
Pin 4	PORTA.2	A-to-D conversions	
Pin 5	PORTA.3	5K ohm Potentiometer 2	Clock chips U6
Pin 6	PORTA.4	This specific pin has special pull-up needs!	No analog function
Pin 7	PORTA.5	A-to-D conversion	Memory chips
Pin 8	PORTE.0	LCD writing controls	
Pin 9	PORTE.1	LCD writing controls	
Pin 10	PORTE.2	LCD writing controls	Communications
Pin 11	Vdd	Logic power	Has no other use
Pin 12	Vss	Logic ground	Has no other use
Pin 13	OSC1	Oscillator	Has no other use
Pin 14	OSC2	Oscillator	Has no other use
Pin 15	PORTC.0	Servo/clock	
Pin 16	PORTC.1	Clock chips	Memory chips, servo/clock, HPWM
Pin 17	PORTC.2	Piezo speaker	HPWM
Pin 18	PORTC.3	Clock chips	Memory chips, servo/clock
Pin 19	PORTD.0	LCD and LED bargraph	
Pin 20	PORTD.1	LCD and LED bargraph	
Pin 21	PORTD.2	LCD and LED bargraph	
Pin 22	PORTD.3	LCD and LED bargraph	
Pin 23	PORTC.4	Used with Memory chips	

Pin 24	PORTC.5	Clock chips	A-to-D conversion, memory chips
Pin 25	PORTC.6	Transmit serial communications	RS232C
Pin 26	PORTC.7	Receive serial communications	RS232C
Pin 27	PORTD.4	LCD and LED bargraph	
Pin 28	PORTD.5	LCD and LED bargraph	
Pin 29	PORTD.6	LCD and LED bargraph	
Pin 30	PORTD.7	LCD and LED bargraph	
Pin 31	Vss	Logic ground	Has no other use
Pin 32	Vdd	Logic power	Has no other use
Pin 33	PORTB.0		Keypad inputs
Pin 34	PORTB.1		Keypad inputs
Pin 35	PORTB.2		Keypad inputs
Pin 36	PORTB.3	Progr'g device	Keypad inputs
Pin 37	PORTB.4		Keypad inputs
Pin 38	PORTB.5		Keypad inputs
Pin 39	PORTB.6	Progr'g device	Keypad inputs
Pin 40	PORTB.7	Progr'g device	Keypad inputs

PORTB lines set as inputs can be pulled up internally with a software instruction. Interrupts can be generated by changes on PORTB lines when they are programmed to do so.

USING THE A-TO-D CAPABILITIES OF THE PIC 16F877A

You can make a number of basic measurements with the LAB-X1 board by using its analog-to-digital conversion capabilities. The resolution of the conversion can be 8 or 10 bits. Still higher resolutions are available if we use ICs that go in empty socket U6. The measurements we make can be used to determine the following:

- Resistance
- Capacitance
- Voltage
- Frequency (This is, of course, not an A-to-D function.)

Resistance is determined by measuring how long it takes a resistor to discharge a capacitor that has just been charged. The measurement is as accurate as the value of the capacitor. The measurement parameters may need to be adjusted in real time to get a

PERIPHERAL FEATURES

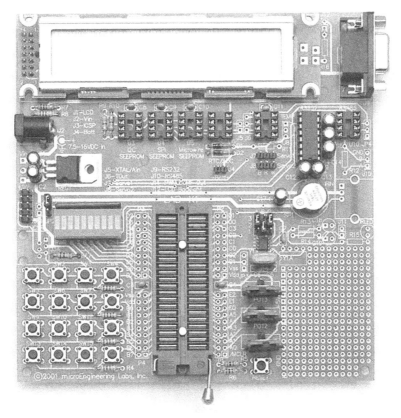

Figure 3.1 A nearly full-size image of the versatile LAB-X1 experimental board.

usable reading (meaning the value of the two components must be selected to get a reading in a reasonable time with reasonable accuracy).

If the relative position of the wiper on a variable potentiometer is required, the A-to-D conversion capabilities of the LAB-X1 can be used to read the potentiometer wiper position (not the resistance). The A-to-D converter always measures the voltage across the device that you connect to the analog input port. You have the choice of reading the value to a resolution of either 8 or 10 bits. If you are reading an 8-bit A-to-D value, the value across the resistance is divided into 256 divisions and the reading will always be between 0 and 255 (inclusive). If you are doing a 10-bit A-to-D conversion, the value will be between 0 and 1023 (inclusive), but since 1 byte can hold only 8 bits, the remaining 2 bits must be read from another register. This is explained in more detail in the section on setting up A-to-D conversions for the IC in socket U6. Here, let's take a quick look at the POT command (used to read a potentiometer) to see how this works.

The POT Command The compiler provides the POT command to make it easy to read the resistive load placed on a pin. In order to use this command, it is necessary to set up the connection to the LAB-X1 as follows:

Set up the pin used on the MCU for analog mode

Select the pin to be used for input

Select what the excitation voltage source will be (internal or external)

On the 16F877A, only 16 pins can be used with the POT command, and these 16 pins have been assigned the aliases from PIN0 to PIN15 in the include file BS1DEFS.BAS (see the PBP manual). For the 16F877A, these are the pins on PORTB (0 to 7) and PORTC (8 to 15). Other PIC MCU have different designations based on their pin counts. See page 26 of the PICBASIC PRO manual for more information.

The POT command is:

`POT Pin, Scale, NMBR`

Where:

Pin is the pin number we have been discussing

Scale is the adjustment for various RC constants. If the RC constant is large, the value of Scale should be small. Scale is determined experimentally with a potentiometer in place of the resistive load. At the low end of the resistance, the value of scale should be 0, and at the high end it should be 255.

NMBR is the variable the result will be placed in.

Values between 5 and 50K ohms may be read with a 0.1 uF capacitor, as shown in the Compiler manual under the POT command.

Capacitance Capacitance can be measured by determining how long it takes to charge a capacitor through an accurately calibrated resistor or by setting up an oscillator with the two components and measuring the oscillator frequency.

Voltage Voltage is measured by setting up an appropriate dividing network with precision resistors and measuring the voltage across an appropriate resistance.

Frequency The PIC 16F877A can measure frequencies directly (as made possible by the compiler software). The timers and counters within the MCU are used to set the measurement intervals and counting hardware.

Reading Switches Switches can be read from the lines of any port set up as an input port. De-bouncing must be performed either in hardware or in software to avoid false readings. (The BUTTON command provides flexible de-bouncing capabilities.)

Make sure that other hardware that may be connected to the pins does not interfere with the switch function and its detection.

Reading Switches in a Matrix Switches arranged in a matrix can be read by setting and reading the rows and columns in the matrix. The technique activates one row of buttons at a time by making it high or low and then seeing if any of the columns have been affected. A detailed description of how this is done is in Chapter 5.

Configuring and Controlling the Properties of the Ports

The PIC 16F877A provides 33 I/O pins distributed across five ports. Each of the ports has unique capabilities built into it. The following section of this book discusses the capabilities of each of the ports with special attention to these properties.

These descriptions are cursory and are designed to provide a quick and ready reference. Refer to the actual datasheet for detailed information on these ports. The datasheet provides information at a level that cannot be provided in a short introductory text like this. Referring to the datasheets frequently will make you comfortable with them in a short time.

PORTA

PORTA is a 6-bit-wide bidirectional port with both analog and digital capability.

Note *The general rule is that if a PIC device has any analog inputs built into it, it will come up as an analog input device on reset and/or startup.*

The PIC 16F877A has analog capability on PORTA (and PORTE) so it comes up as an analog device on startup and reset. If you are going to use it as a digital device, you must set register ADCON as needed. The most common value is *ADCON1 to %00000111* for using the LCD display. This line of code will be seen in many of the programs in this book and is explained in Chapter 5. (%00000111 sets all the analog pins to digital, but there are many other choices.)

The PIC 16F877A supports external access to only six of the eight pins on this port. Each of the six pins may be set to function as an input or output by appropriately loading the TRISA register. A "0" (zero) in a bit in this register sets the corresponding pin to function as an output, while a "1" (one) sets it to function as an input.

Thus, setting...

```
TRISA=%00111000
```

(% means this is a binary number. We will use this binary notation throughout this book because it makes it easier to see what each bit is being set to. Bit 7, the most significant bit, is on the left; bit 0, the least significant, is on the right.)

Would make lines A0, A1, and A2 outputs and lines A3, A4, and A5 inputs. The most *significant 2 bits are ignored* (and could be set to 1s or 0s) because PORTA has only six active lines. (However, the 2 ignored bits are used by the processor and they can be read when necessary. You should not set them. Again, see the datasheet for specific details.)

As mentioned earlier, the specific functions (the uses the pins are put to) of the pins are controlled with the ADCON1 (the first **A-to-D CON**trol) register.

All the pins have TTL level inputs and full CMOS level output drivers. This makes it easy to connect these lines directly to standard logic components (meaning that *usually* no intermediate resistors are needed between components if TTL or CMOS components are interconnected).

PORTA designations are somewhat complicated. Pins A0, A1, A2, A3, (skip A4), and Pin A5 can be configured as analog inputs by setting the ADCON1 register appropriately. Pin A3 is also used as a voltage input for comparing with the analog voltage inputs on other pins. Pin PORTA4, is used as input for the timer TIMER0. It is called T0CK1 for Timer0 clock Input. This is Pin 6 of the PIC and is used as the input pin for TIMER0 only when configured as such. It is a Schmitt trigger input with open drain output. Open drain means it acts like the contacts of a tiny relay. Schmitt triggered inputs have increased noise immunity (because of the hysteresis between state changes).

The two registers that control PORTA are TRISA and ADCON1.

ADCON1 controls the A-to-D and voltage reference functions of PORTA. The setting of the various bits select a complicated set of conditions that are described in detail in a table in the datasheet and in this book. (In the preceding discussion, when ADCON1 was set to %00000111, we were accessing this feature.)

Pin A4 has special needs when used as an output. It can be pulled down low but will float when set high. It must be pulled up with a (10K to 100K) resistor to tie it high when this is necessary. This pin has an open drain output rather than the usual bipolar state of the other pins. As noted earlier, this pin is skipped in the A-to-D conversion table. (Ignoring this fact is a source of many common mistakes, so keep this in mind.)

PORTB

PORTB is a full 8-bit-wide bidirectional port.

Internal circuitry (meaning built into the MCU) allows all the pins on PORTB to be pulled up to a high state (very weakly) by setting pin 7 of the option register (OPTION_REG.7) to 0. Only those pins designated at inputs are affected. These pull-ups are disabled on startup and on reset.

Pins B3, B6, and B7 are used for the low voltage programming of the PIC. Bit 3 in TRISB must be cleared (set to 0 or pulled down to 0) to negate the pull up on this pin to allow programming to take place. See pages 42 and 142 in the datasheet for more information regarding the B3 pin. *This is very important.*

Pins B4 to B7 will cause an interrupt to occur when their state changes if they are configured as inputs and the appropriate interrupts are configured. Any of these four pins that are configured as outputs will be excluded from the interrupt feature. The interrupts are controlled by the INTCON (interrupt control) register. This PORTB interrupt capability has the special feature of letting it be used to awaken a sleeping MCU.

Pin B0 has special interrupt functions that are controlled through the INTEDG bit, which is bit 6 of the OPTION_REG. See datasheet for more information. *External interrupts are routed to the PIC through this pin.* External interrupts are those that are created by events outside the PIC MCU.

The three registers that control PORTB are TRISB, INTCON, and OPTION_REG.
OPTION_REG controls the optional functions of PORTB as follows.

- Bit B0 of OPTION_REG sets prescalar value
- Bit B1 of OPTION_REG sets prescalar value
- Bit B2 of OPTION_REG sets prescalar value
- Bit B3 of OPTION_REG sets the prescalar option (used in low-voltage programming)
- Bit B4 of OPTION_REG sets Timer0 input pulse edge selection
- Bit B5 of OPTION_REG sets the clock selection
- Bit B6 of OPTION_REG sets edge selection for interrupts (programming uses)
- Bit B7 of OPTION_REG sets the pull-ups when cleared (programming uses)

PORTC

PORTC is a full 8-bit-wide bidirectional port.

All the pins on PORTC have Schmitt trigger input buffers. This means they are designed to be more immune to noise on these input lines.

The alternate function of the PORTC pins are defined as follows:

- Bit C0   I/O pin or Timer1 Oscillator output or Timer1 Clock input
- Bit C1   I/O pin or Timer1 Oscillator input or Capture 2 input or Compare 2 output or Hardware PWM2 output
- Bit C2   I/O pin or Capture 1 input or Compare 1 output or Hardware PWM1 output
- Bit C3   I/O pin or Synchronous clock for both SPI and I2C memory modes
- Bit C4   I/O pin or SPI data or data I/O for I2C mode
- Bit C5   I/O pin or Synchronous serial port data output
- Bit C6   I/O pin or USART Asynchronous transmit or synchronous clock
- Bit C7   I/O pin or USART Asynchronous receive or synchronous data

Special care must be taken when using PORTC's special function capabilities, in that some of these functions will change/set the I/O status of certain other pins when in use, and this can cause unforeseen complications in the function of other capabilities. See the datasheet for details.

The register that controls PORTC is the TRISC register. No other registers are involved. DEFINEs are used to set or control certain PORTC functions.

Note   *The speaker on the LAB-X1 board is connected to pin C.1, so the use of this pin is limited because the noise generated by the speaker when this pin is used can be very irritating to humans. Since this is one of the lines that allows the generation of continuous background PWM signals (HPWM 2), it compromises the clean use of this pin unless the speaker is removed or the signal is put through a gate to clean it up. (I recommend that you avoid modifications to the board if you can.) The load of the tiny speaker loads the pin and can compromise certain other uses, but in most cases you can* *use this pin without concern. (Under certain situations, the frequency on this pin can be set high enough to be above human hearing range, allowing the pin to be used like the others.)*

PORTD

PORTD is a full 8-bit-wide bidirectional port.

All the pins on PORTD have Schmitt trigger input buffers. This means they are designed to be more immune to noise on the input lines.

PORTD can also be configured as a microprocessor port by setting PSPMODE through setting TRISE.4 to 1. (Note that we are specifying bit 4 of PORTE here internally; there is no external pin 4.) In this mode, all the input pins are in TTL mode. (See page 48 of the datasheet.)

The alternate function of the PORTD pins are defined as follows:

Bit D0 or parallel slave port bit 0

Bit D1 or parallel slave port bit 1

Bit D2 or parallel slave port bit 2

Bit D3 or parallel slave port bit 3   Port D is the

Bit D4 or parallel slave port bit 4   parallel slave port

Bit D5 or parallel slave port bit 5

Bit D6 or parallel slave port bit 6

Bit D7 or parallel slave port bit 7

The registers that control PORTD are TRISD and TRISE. TRISE controls the operation of the PORTD parallel slave port mode when Bit TRISE.4 is set to 1. (Remember again that only pins E0, E1, and E2 are available external to the MCU on PORTE.)

Slave port functions as set by PORTE when Bit TRISE.4 is set to 1 are:

Bit TRISE.0 direction control of Pin PORTE.0 / RD / AN5

Bit TRISE.1 direction control of Pin PORTE.1 / WR / AN6

Bit TRISE.2 direction control of Pin PORTE.2 / CS / AN7

Bit TRISE.3. NOT USED

Bit TRISE.4 Slave port select   1=Slave port selected

                                         0=Use as standard I/O port

Bit TRISE.5 Buffer overflow detect   1=Write occurred before reading old data

                                         0=No error occurred

Bit TRISE.6 Buffer status   1=Still holds word

                                  0=Has been read

Bit TRISE.7 Input buffer status   1=Full

                                  0=Nothing received

Read the datasheet a couple of times to get a better understanding of these operations. The preceding is a very brief overview and is intended only to alert you and give you an idea of what the possibilities are.

PORTE

PORTE is only 3 external bits wide and is a bidirectional port. The other bits are internal and are used as mentioned earlier under PORTD (to which they are related). All external three pins can also be used as analog inputs.

All the pins on PORTE have Schmitt trigger input buffers when used as digital inputs. The alternate function of the PORTE pins are defined as follows:

Pin RE0 direction control of Pin PORTE.0 / RD / AN5

Pin RE1 direction control of Pin PORTE.1 / WR / AN6

Pin RE2 direction control of Pin PORTE.2 / CS / AN7

TIMERS

The PIC 16F877A MCU has four timers: a watchdog timer and three regular timers. These allow the accurate timing (and counting) of chronological events. Timers are discussed in much greater detail in Chapter 6.

Some of the timers have (pre- and post-) scalars associated with them that can be used to multiply the timer setting by an integer amount. As you can imagine, the scaling ability is not adequate to allow all timed intervals to be created. We also must consider the uncertainty in the frequency of the clocking crystal, which is usually not exactly what it is stated to be (and may drift with ambient temperature). This means that though fairly accurate timings can be achieved with the hardware as received, additional software adjustments may have to be added if more accurate results are desired. This is done by having the software make a correction to the time every so often. (This also means that an external source, that is at least as accurate as the result we want, is needed to verify the timing accuracy of the device created.)

As always, the three timers in the microcontroller are clocked at a fourth of the oscillator speed, meaning that a timer within a 4 MHz system clock gets a counting signal at 1 MHz.

Very simply stated, an 8-bit timer will count up to 255 and then flip to zero and start counting from 0 to 255 again. An interrupt occurs (a bit is set) every time the timer register overflows from 255 to 0. We respond to the interrupt by doing whatever needs to be done in response to the interrupt and then resetting the interrupt flag bit (to 0). On timers that permit the use of a prescalar, the prescalar lets us increase the time between interrupts by multiplying the time between interrupts by a definable value in a 2-, 3-, or 4-bit location. On timers that can be written to, we can start the counter wherever we like (to change the interrupt timing), and on timers that can be read, we can read the contents whenever we want to.

Example: A 1-second timer setting with a prescalar set to 16 would provide us with an interrupt every 16 seconds, and we would have 16 seconds to do whatever we wanted to do between the interrupts or we would miss the next interrupt.

If we needed an interrupt every 14.5 seconds, we would use a timer set to 0.5 seconds and a prescalar of 29 if 29 was specifiable (which it is not). So not all time intervals can be created with this strategy because there are limits as to what can be put in the timer and what can be put in the prescalar when we are using 8/16-bit registers and specific oscillator speeds.

WATCHDOG TIMER

A watchdog timer is a timer that sets an interrupt when it runs out so as to tell us the program has hung up or gone awry for some reason. As such, it is expected that in a properly written program, the watchdog timer will never set an interrupt. This is accomplished by resetting the watchdog timer every so often within the program. The compiler does this automatically if the watchdog timer option is set. Setting the option does not guarantee a program that cannot hang up. Software errors and infinite loops that reset the timer within them can still cause hangups.

PRESCALARS AND POSTSCALARS

The value of the scaling factor that will be applied to the timer is determined by the contents of 2 or 3 bits in the interrupt control register. These bits multiply the time between interrupts often by powers of two. Prescalars and postscalars have the same effect on the interrupts: They delay them.

COUNTERS

Only Timer0 and Timer1 can be used as counters. Timer2 cannot be used as a counter because it does not have an external input line. Both the timers and the counters are covered in detail in Chapter 6.

THE SOFTWARE, THE COMPILERS, AND THE EDITOR

MicroEngineering Labs, Inc. *provides two BASIC compilers* that make writing the code for the PIC family of microcontrollers offered by Microchip Technology tremendously easy. In this book, we will discuss the more powerful of the two: the PICBASIC PRO Compiler. A listing of the commands provided by each compiler is provided in the following to let you *compare the two* compilers and select the one best suited to both your budget and your needs.

The Basic Compiler Instruction Set

The following is the smaller compiler of the two:

ASM..ENDASM	Insert assembly language code section.
BRANCH	Computed GOTO (equivalent to ON..GOTO).
BUTTON	De-bounce and auto-repeat input on specified pin.
CALL	Call assembly language subroutine.
EEPROM	Define initial contents of on-chip EEPROM.
END	Stop execution and enter low power mode.
FOR..NEXT	Repeatedly execute statement(s).
GOSUB	Call BASIC subroutine at specified label.
GOTO	Continue execution at specified label.

HIGH	Make pin output high.
I2CIN	Read bytes from I2C device.
I2COUT	Send bytes to I2C device.
IF..THEN - GOTO	If specified condition is true.
INPUT	Make pin an input.
[LET]	Assign result of an expression to a variable.
LOOKDOWN	Search table for value.
LOOKUP	Fetch value from table.
LOW	Make pin output low.
NAP	Power down processor for short period of time.
OUTPUT	Make pin an output.
PAUSE	Delay (1 msec resolution).
PEEK	Read byte from register.
POKE	Write byte to register.
POT	Read potentiometer on specified pin.
PULSIN	Measure pulse width (10 μsec resolution).
PULSOUT	Generate pulse (10 μsec resolution).
PWM	Output pulse width modulated pulse train to pin.
RANDOM	Generate pseudorandom number.
READ	Read byte from on-chip EEPROM.
RETURN	Continue execution at statement following last executed GOSUB call.
REVERSE	Make output pin an input or an input pin an output.
SERIN	Asynchronous serial input (8N1).
SEROUT	Asynchronous serial output (8N1).
SLEEP	Power down processor for a period of time (1 sec resolution).
SOUND	Generate tone or white noise on specified pin.
TOGGLE	Make pin output and toggle state.
WRITE	Write byte to on-chip EEPROM.

MATH OPERATIONS

All math operations are unsigned and performed with 16-bit precision:

+	Addition
–	Subtraction
*	Multiplication
**	MSB of multiplication
/	Division
//	Remainder
MIN	Minimum
MAX	Maximum
&	Bitwise AND
|	Bitwise OR
^	Bitwise XOR
&/	Bitwise AND NOT
|/	Bitwise OR NOT
^/	Bitwise XOR NOT

The PICBASIC PRO Compiler Instruction Set

This is the larger compiler of the two:

@	Insert one line of assembly language code.
ADCIN	Read on-chip analog to digital converter.
ASM..ENDASM	Insert assembly language code section.
BRANCH	Computed GOTO (equivalent to ON..GOTO).
BRANCHL	Branch out of page (long BRANCH).
BUTTON	De-bounce and auto-repeat input on specified pin.
CALL	Call assembly language subroutine.
CLEAR	Zero all variables.

CLEARWDT	Clear (tickle) watchdog timer.
COUNT	Count number of pulses on a pin.
DATA	Define initial contents of on-chip EEPROM.
DEBUG	Asynchronous serial output to fixed pin and baud.
DEBUGIN	Asynchronous serial input from fixed pin and baud.
DISABLE	Disable ON DEBUG and ON INTERRUPT processing.
DISABLE DEBUG	Disable ON DEBUG processing.
DISABLE INTERRUPT	Disable ON INTERRUPT processing.
DTMFOUT	Produce touch-tones on a pin.
EEPROM	Define initial contents of on-chip EEPROM.
ENABLE	Enable ON DEBUG and ON INTERRUPT processing.
ENABLE DEBUG	Enable ON DEBUG processing.
ENABLE INTERRUPT	Enable ON INTERRUPT processing.
END	Stop execution and enter low power mode.
ERASECODE	Erase block of code memory.
FOR ..NEXT	Repeatedly execute statements.
FREQOUT	Produce up to two frequencies on a pin.
GOSUB	Call BASIC subroutine at specified label.
GOTO	Continue execution at specified label.
HIGH	Make pin output high.
HPWM	Output hardware pulse width modulated pulse train.
HSERIN	Hardware asynchronous serial input.
HSERIN2	Hardware asynchronous serial input, second port.
HSEROUT	Hardware asynchronous serial output.
HSEROUT2	Hardware asynchronous serial output, second port.
I2CREAD	Read from I2C device.
I2CWRITE	Write to I2C device.
IF..THEN..ELSE..ENDIF	Conditionally execute statements.
INPUT	Make pin an input.
LCDIN	Read from LCD RAM.

LCDOUT	Display characters on LCD.
{LET}	Assign result of an expression to a variable.
LOOKDOWN	Search constant table for value.
LOOKDOWN2	Search constant / variable table for value.
LOOKUP	Fetch constant value from table.
LOOKUP2	Fetch constant / variable value from table.
LOW	Make pin output low.
NAP	Power down processor for short period of time.
ON DEBUG	Execute BASIC debug monitor.
ON INTERRUPT	Execute BASIC subroutine on an interrupt.
OWIN	One-wire input.
OWOUT	One-wire output.
OUTPUT	Make pin an output.
PAUSE	Delay (1 msec resolution).
PAUSEUS	Delay (1 μsec resolution).
PEEK	Read byte from register.
PEEKCODE	Read byte from code space
POKE	Write byte to register.
POKECODE	Write to code space at device programming time.
POT	Read potentiometer on specified pin.
PULSIN	Measure pulse width on a pin.
PULSOUT	Generate pulse to a pin.
PWM	Output pulse width modulated pulse train to pin.
RANDOM	Generate pseudorandom number.
RCTIME	Measure pulse width on a pin.
READ	Read byte from on-chip EEPROM.
READCODE	Read word from code memory.
REPEAT..UNTIL	Execute statements until condition is true.
RESUME	Continue execution after interrupt handling.
RETURN	Continue at statement following last GOSUB call.

REVERSE	Make output pin an input, or an input pin an output.
SELECT CASE	Compare a variable with different values.
SERIN	Asynchronous serial input (BS1 style).
SERIN2	Asynchronous serial input (BS2 style).
SEROUT	Asynchronous serial output (BS1 style).
SEROUT2	Asynchronous serial output (BS2 style).
SHIFTIN	Synchronous serial input.
SHIFTOUT	Synchronous serial output.
SLEEP	Power down processor for a period of time.
SOUND	Generate tone or white noise on specified pin.
STOP	Stop program execution.
SWAP	Exchange the values of two variables.
TOGGLE	Make pin output and toggle state.
USBIN	USB input.
USBINIT	Initialize USB.
USBOUT	USB output.
WHILE..WEND	Execute statements while condition is true.
WRITE	Write byte to on-chip EEPROM.
WRITECODE	Write word to code memory.
XIN	X-10 input.
XOUT	X-10 output.

MATH FUNCTIONS / OPERATORS

The math operations are unsigned and performed with 16-bit precision:

+	Addition
−	Subtraction
*	Multiplication
**	Top 16 bits of multiplication
*/	Middle 16 bits of multiplication
/	Division

//	Remainder (modulus)
<<	Shift left
>>	Shift right
ABS	Absolute value
COS	Cosine
DCD	2n decode
DIG	Digit
DIV32	31-bit × 15-bit divide
MAX	Maximum
MIN	Minimum
NCD	Encode
REV	Reverse bits
SIN	Sine
SQR	Square root
&	Bitwise AND
|	Bitwise OR
^	Bitwise exclusive OR
~	Bitwise NOT
&/	Bitwise NOT AND
|/	Bitwise NOT OR
^/	Bitwise NOT exclusive OR

As can be seen from the preceding comparison, the *PICBASIC PRO Compiler provides a much more comprehensive instruction set* and therefore is the compiler of choice for serious development work.

It is, of course, also possible to program microcontrollers in assembly language and "C," but this book does not cover these languages. A number of good books are available on the subject, and some that I looked over are listed in a file on the support Web site with my comments. Some educators feel that a Junior College level class on the subject is the best way to learn how to do this and there is some merit to this but for our purposes the PICBASIC PRO Compiler will do everything we need and is much easier to use.

In addition to the compiler, *you need an editor* to allow you to write and edit programs with ease. A very adequate editor is provided as a part of the compiler package. Its called the MicroCode Studio editor. This comprehensive and powerful editor is also available at no charge (on the Internet) from MicroCode Studios. This is a complete editor with no limit on the number of lines of code you can write. It is fully integrated with the software and hardware provided by microEngineering Labs and is the editor of choice for most users. (The free version is limited to compiling programs for just a few microcontrollers, but these include both the *16F877A* and the *16F84A*.)

Three of the editors available are:

- **MicroCode Studio** Mecanique's MicroCode Studio is a powerful, visual, integrated development environment (IDE) with an In-Circuit Debugging (ICD) capability designed specifically for microEngineering Labs' PICBASIC PRO Compiler. This software can be downloaded from the Internet at no charge. The only limitation on the software is that it allows you to run only one IDE at one time. This is not a real handicap at our level of interest, however, since it is the editor that best suits our needs, and because all programs in this book were written with this editor.
- **Proton+ Lite BASIC editor** (provided by Crownhill) This is a test version of their editor and is limited to 50 lines of code and three processors (including the PIC 16F877A). If you like this editor, you can use this as your main editor and then cut and paste to the MicroCode Studio to compile and run your programs, and thus go around the 50-line limit. The *native language of this editor is not the same as* PICBASIC, so there are other handicaps to contend with. (These incompatibilities are best avoided.)
- **Microchip MPLAB** This is the software that the maker of the microcontrollers, Microchip Technology Inc., provides for editing programs written for their PIC series of microcontrollers. It is an assembly level programmer. We will not be doing any assembly language programming, but the editor can be useful.

PICBASIC PRO Compiler

The PICBASIC PRO Compiler—hereafter referred to as the PBP—provides all the functions needed to program almost the entire family of PIC microcontrollers in a BASIC-like environment. This means it allows you to write programs that read the inputs and write to the outputs in a simple and easy-to-learn fashion. It means that communications are simplified and that the time it takes to get an application running is reduced manyfold. It also means that the programs are easier to follow and debug (though debugging can get quite complicated even on these seemingly simple devices). The compiler supports only integer math, but that is not a big handicap when we are working with these limited microprocessors.

It also means that the programs that are developed are shorter than assembly language programs, but they are slower in their execution than assembly language programs. (Complications also exist that have to do with the use of interrupts, and which must be addressed, but these are beyond the scope of this book.)

All the exercises and examples provided in the text are based on the PBP compiler. We will *not* go over the detailed instructions for using each of the PBP instructions in the text. It is expected and will be assumed that you will have purchased the software and thus will have the manual for the compiler in hand. However, some commands can be complicated to implement and so we will spend time on them as necessary.

The compiler is kept current by microEngineering Labs. for the latest MCUs released by Microchip Technology. The LAB-X1 uses the 16F877A MCU, and it is the MCU of choice, though other MCUs that have a general pin-for-pin compatibility with this MCU may also be used. All the experiments and exercises in this book will use the PIC 16F877A only. The compiler addresses almost all the capabilities of this MCU, and we will cover the use of all the devices provided on the LAB-X1 board to develop the comfort level you need to incorporate them into your instruments and controllers.

Detailed instructions for installing the software on your PC are provided in the compiler manual. It is not necessary to install the software from a DOS prompt. It is much easier to install it under Windows with the "Install.exe" or equivalent file provided in each package.

The software can be set up so that "one mouse click" will transfer the program from the editor to the PIC microcontroller and run the program in the PIC. In order to do this, you have to add a couple of functional codes to the programmer operating system. These codes tell the programmer to load the program and execute it. Installing the software is covered in detail in the chapter on getting started.

A SIMPLE EXAMPLE PROGRAM USING PICBASIC

A program that makes the LEDs blink ON and OFF is usually the first program written by beginners. The purpose of the program is not to blink the LEDs but rather to allow you to go through the programming procedures in a simple and straightforward way, and get a result that is easy to verify. Once you have the LEDs blinking, you will know you have followed all the steps necessary to write and execute a program. Larger, more complicated programs may be much more difficult to write and debug but they are no more difficult to compile, load, and run.

Program 4.1 outlines the keystrokes for writing and running the "blink" the LEDs program in PICBASIC.

Program 4.1 The first program (Blinking all eight LEDs on PORTD one at a time)

```
; ******************************************************************
; *   Name        myBlink8leds.BAS
; *   Author      Harprit Singh Sandhu
; *   Notice      Copyright © 2008
; *               All Rights Reserved
; *   Date        1/Feb/2008
; *   Version     1.0
; *   Notes       Blinks all 8 LEDs on bargraph one at a time
;
; ******************************************************************
CLEAR                       ; clear RAM memory
DEFINE OSC 4                ; define the osc freq
LED_ID VAR BYTE             ; call out the two variables LED_ID and I
I VAR BYTE                  ; as 8 bit bytes
```

(Continued)

Program 4.1 The first program (Blinking all eight LEDs on PORTD one at a time) (*Continued*)

```
TRISD =%00000000          ; set PORTD to all outputs
                          ;
MAINLOOP:                 ; loop is executed forever
  I=1                     ; initialize the counter to 1
  FOR LED_ID = 1 TO 8     ; do it for the 8 LEDs
    PORTD=I               ; puts number in PORTD
    PAUSE 100             ; pause so you can see the display
    I=I * 2               ; multiplying by 2 moves lit LED left 1 pos
  NEXT LED_ID             ; go up and increment counter
GOTO MAINLOOP             ; do it all forever
END                       ; always end with END statement
```

PICBASIC PRO TIPS AND CAUTIONS

1. To get context-sensitive help, move the cursor over a PICBASIC command, click to set the cursor, and press F1.

2. Programs assume the PIC is running at 4 MHz. To change the default setting (for example, to 20 MHz), simply add DEFINE OSC 20 at the top of your program and set the oscillator jumpers on the LAB-X1 accordingly. It is good practice to always specify the oscillator speed in a program. Beginners should start with 4 MHz designs. The LAB-X1 is set up to run at 4 MHz as received from the factory. See the manual. The defined OSC speed has to match the hardware crystal for the hardware and software to work correctly. This is especially important for time-sensitive instructions.

3. Before you can use the LCDisplay on the LAB-X1, ADCON1 must be set (to %00000111 to make PORTE [and PORTA] digital) and you must pause about 500 msec to allow the LCD to start up before issuing it its first command. You may not need a pause, or a shorter pause may be specified, if there are many time-consuming instructions before the first LCDOUT instruction is executed. (Other values of ADCON1 can also be used, depending on how you want the A and E ports configured. See the discussion in Chapter 18.)

4. I have used binary notation (%01010101) throughout this book to set relevant bytes and registers so you can readily see which bit is being set to what. The compiler accepts hexadecimal and decimal notation just as willingly. Binary notation does not permit a space after the % sign, and all eight bits must be specified.

5. When using Word for programming support, if a single quotation mark (') is copied from a Word file and pasted into the MicroCode Studio editor, it will be interpreted as a (`) and will therefore *not properly start the comment section* of the line. All these have to be changed in the editor after pasting. Pasting from the editors into Word does not exhibit the same effect. If you use (;) for the comments, this problem does not occur.

6. All the named registers can be called by name when using the compiler. The register names are the same as those used (defined) by the manufacturer in the datasheet and are the same across the entire family of PIC microcontrollers if they provide the same function. Uppercase or lowercase names can be used. The DEFINEs must be

stated in uppercase only, and the spellings in the DEFINE lines are not always checked by the compiler! Therefore, be very careful with the spellings when adding DEFINEs to your program.
7. Circuits and segments of circuits are provided throughout this book to show you how to connect up the hardware when you design your own circuits. If you have access to AutoCAD, you can cut and paste the diagrams in the files on the support Web site into your own designs. All the diagrams in this book are on the Web site.

A FREE DEMO BASIC COMPILER

A free version of the PICBASIC PRO Compiler by microEngineering Labs can also be downloaded from the microEngineering Labs Web site. This is a fully functional compiler with the limitation that programs are limited to 30 lines of code. This is enough to allow you to test the compiler and any instruction that you might have a special interest in. This version can give you a good idea of the power and ease of use of the language. I encourage you to try out this compiler before you make your compiler purchase.

5
CONTROLLING THE OUTPUT
AND READING THE INPUT

General

After all is said and done, it's about the input and output, and what happens between the two of them. It has to do with how we use the many capabilities provided within each PIC/MCU, and how creative we are about using those capabilities. In this chapter, we start learning to use all the input and output capabilities provided on the 16F877A using LAB-X1.

All the programs we will be discussing are provided on the support Web site (www.encodergeek.com), which is maintained as part of this book. Simply navigate to your area of interest on the Web site.

You can copy the files from the Web site and run them on the 16F877A in the LAB-X1. The exercises listed in each chapter are exercises that are designed to increase your familiarity and competence with the 16F877A. The answers to them are not provided.

In preparation for writing programs, set up the LAB-X1 so it can be programmed with one mouse button click or by pressing F10, as described in detail in Appendix A of this book.

The I/O that uses ICs in the seven empty sockets on the LAB-X1 board is covered separately in Chapters 7 and 8. These chapters also cover one-wire memory, A-to-D converters, and a number of related thermometric devices.

The I/O that uses the serial port (as RS-232 or RS485) is covered in Chapter 8. This covers communications between PICs and personal computers.

We will learn about input and output by first writing simple programs that control the outputs, and then write programs that read the inputs. We will learn how to control the outputs first, because this can be done directly from the software without need for any input. Once we control the output, we will learn how to read the inputs and make the outputs respond to them in a useful and coherent manner. The following is a list of the programs to be developed.

Programs That Create Output

The output is what we are looking for in any instrument. It is the final result of all the work we do. We will interact with all of the output devices on the LAB-X1 in these programs.

BASIC PROGRAMS

- Write a program to control one LED on the bargraph.
- Control all eight LEDs in the bargraph consecutively.
- Dim and brighten one LED (creating pulses of various lengths).
- Write "Hello World" to the LCD on its two lines.
- Write binary and decimal values to the LCD.
- Output a simple tone on the speaker.
- Output a telephone tone signal on the speaker.

ADVANCED PROGRAM

- Move an R/C servo back and forth

Programs That Read the Inputs and Then Provide Output

1. Write a simple program to read the first column, first row button, and turn ON one LED while the button is down.
2. Read the entire keyboard and display the binary value of the row and column read on the LCD.
3. Read the keyboard and display decimal key number on the LCD.
4. Read one potentiometer and display its 8-bit value on the LCD in binary, hex, and decimal notation. Also display the binary value on the bargraph.
5. Read all three potentiometers and display their values on the LCD.
6. Advanced: Use the three potentiometers to control an R/C servo. Control the location of the center position, the limit position of the end positions, and the rate of movement. Use three switches on the keyboard to move the servo clockwise, center the servo, and move it counterclockwise.

Creating Outputs

It will be easier if we learn to control the outputs first because we can do this with programs that we write without the need for any additional hardware or input signal. We will start with the simple control of LEDs, proceed to the control of the two-line LCD,

CREATING OUTPUTS   49

which is provided on the LAB-X1, and then move on to using the speaker and an R/C hobby servo.

Let's start with the standard "turning an LED ON and OFF" program. We will use one of the LEDs in the ten LED bargraph that is provided on the LAB-X1. We have control of only the eight rightmost LEDs on this bargraph. The leftmost LED is the power ON indicator and the one next to it comes ON if we were using a common cathode arrangement for the bargraph (as opposed to the common anode arrangement as it is currently configured).

The circuitry we are interested in is shown in Figure 5.1. All other circuitry of the LAB-X1 is still in place, but we have suppressed it so we are not distracted by it and can concentrate on the LED that is of interest to us. (PORTD.0). PORTD.0 refers to bit 0 of PORTD.

In our first exercise, we want to control the rightmost LED of the LED array. This is connected to bit 0 of PORTD in the circuitry shown in Figure 5.1. Our program needs to turn this LED ON and OFF to demonstrate that we have complete control of these two functions.

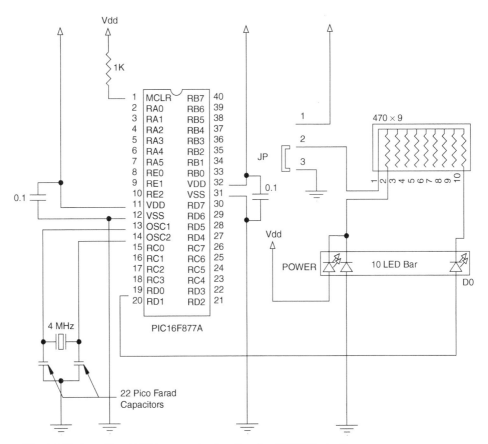

Figure 5.1   The LED bargraph circuitry for PORTD pin 0.

In general, the ports on the microcontrollers (MCUs) are designed so they can be used as inputs or as outputs. In fact, the ports can be programmed so certain pins on a port are inputs and others are outputs. All we have to do is tell the program what we want done and the compiler will handle the details. The compiler not only allows you to define how you will use the pins of each port, it can also set them up as inputs or as outputs automatically, depending on the instructions you use to address the pin in your program. You have a choice of setting PORTD to an output port and then setting pin 1 on this port high, or you can simply tell the compiler to make pin 1 of PORTD high and it will take care of the setting-up details.

The ports can be treated just like any other memory location in the microcontroller. By name, you can read them, set them, and use them in calculations and manipulations just like you can with any other named or unnamed memory location. If things are connected to the ports and pins, the program will interact with, and respond to, whatever is connected to them. (Any named port, register, or pin can be addressed directly by name for all purposes when using the PBP Compiler. They are called out as they are named in the datasheet for each individual PIC by the manufacturer.)

Blink One LED

Type the following program (Program 5.1) into your PC and save it. It does not need to be saved in the same directory as the PBP.exe program. To keep with the conventions being used in the compiler manual, call this program myBLINK.BAS so it does not overwrite the BLINK.BAS program provided on the disk that came with the LAB-X1.

The program to control pin 0 of PORTD is as follows:

Program 5.1 **Controlling (blinking) an LED** (Blinking an LED [We are using the rightmost LED on the bargraph])

```
CLEAR                ; clear memory locations
DEFINE OSC 4         ; osc speed

LOOP:                ; main loop
  HIGH PORTD.0       ; turns LED connected to D0 ON
  PAUSE 500          ; delay 0.5 seconds
  LOW PORTD.0        ; turns LED connected to D0 OFF
  PAUSE 500          ; delay 0.5 seconds
GOTO LOOP            ; go back to Loop and repeat operation
END                  ; all programs must end with END
```

The program demonstrates the most elementary control we have over an output: that of turning it ON and OFF. In this program, we did not have to set the port directions (with the TRIS command), because the HIGH and LOW command take care of that automatically. (If we used PORTD.0=1 instead of HIGH PORTD.0, we would have had to set TRISD to %11111110 first to set all lines to inputs except D0, which is shown as the only pin set to be an output.)

We will use binary notation (%11110000) for setting all pins, ports, port directions, and register values throughout this book, though you can use hexadecimal ($F0) and decimal (DEC 240) notation interchangeably. Using binary notation lets you see what each pin is doing without having to make any conversions.

Blink Eight LEDs in Sequence

In Program 5.2, we blink the eight rightmost LEDs on the bargraph, one LED at a time. The circuitry we are using for this program is shown in Figure 5.2. We do this by setting PORTD to 1 and then multiplying it by two a total of eight times to move the lighted LED left one position in each iteration. Note that the last multiplication overflows the 8-bit counter and turns all the LEDs OFF.

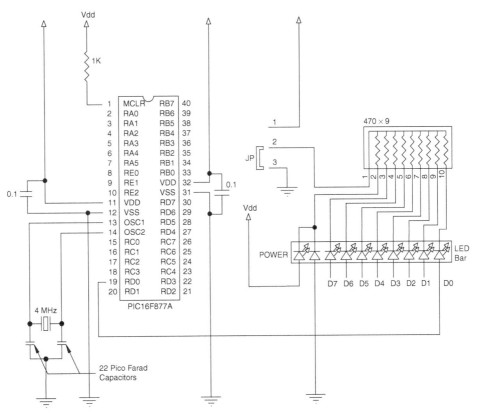

Figure 5.2 The LED bargraph circuitry to all of PORTD.

Program 5.2 Blinking eight LEDs one after the other on a bargraph

```
CLEAR                     ; clear memory
DEFINE OSC 4              ; Osc speed
LED_ID VAR BYTE           ; call out the two variables
A VAR BYTE                ; as 8 bit bytes
TRISD =%00000000          ; set PORTD to all outputs
                          ;
MAINLOOP:                 ; this loop is executed forever
  A=1                     ; initialize the counter to 1
    FOR LED_ID = 1 TO 8   ; do it for the 8 LEDs
      PORTD=A             ; puts number in PORTD
      PAUSE 100           ; pause so you can see the display
      A=A * 2             ; multiply by 2 moves lit LED left 1
                          ; position
    NEXT LED_ID           ; go up and increment counter
GOTO MAINLOOP             ; do it all forever
END                       ; always end with END
```

Dim and Brighten One LED

In Program 5.3, we demonstrate the ability to dim an LED (at PORTD.0) by varying the duty cycle of the ON signal to the LED.

Program 5.3 Turns on an LED and dims the one next to it (Doing it this way lets you compare the brightness of the two LEDs)

```
CLEAR                     ; always start with a CLEAR statement
DEFINE OSC 4              ; osc speed
TRISD = %11111100         ; set only PORTD pin 0 and 1 to outputs
X VAR BYTE                ; declare X as a variable
PORTD.1=1                 ; turned PORTD.1 ON & compare to PORTD.0
                          ;
LOOP:                     ; start of loop
  FOR X = 1 TO 255 STEP 2 ; set up loop for X
    PWM PORTD.0, X, 3     ; vary the duty cycle
    PAUSE 200/X           ; pauses longer for the dimmer values.
  NEXT X                  ; end of loop for X
GOTO LOOP                 ; return and do it again
END                       ; all programs with an END statement
```

With the preceding programs, we learned that we can control the ON-OFF state and the brightness on an LED. Controlling the brightness will become relevant when we are controlling seven segment displays because the LEDs in them are turned ON one at a time and the duty cycle has to be managed properly to get an acceptable display.

LCD Display

These notes describe the use of, and interactions with, existing hardware connections to the liquid crystal display as it comes wired on the LAB-X1 module. These connections are defined in the program with DEFINE statements. Define them as necessary in your designs, keeping in mind that designs that follow the connections for LAB-X1 will allow you to use the LAB-X1 as the test platform. (Even if your design is substantially different from the LAB-X1 circuitry, this will remain a useful feature for testing the LCD and certain I/O that may be the same.)

On the LAB-X1, the LCD data is fed from PORTD, and all 8 bits of this port are connected to the LCD. It is controlled from the 3 bits of PORTD. You therefore have the choice of using only the 4 high bits or PORTD as a 4-bit data path for the LCD or using all 8 bits. The entire port is also connected to eight of the LEDs on the ten-light LED bargraph. (The two leftmost LEDs in the bargraph are used to indicate that the power to the LAB-X1 is ON and the polarity of the bargraph.) The 4 high bits, bits D4 to D7, cannot be used for any other purpose if the LCD is being used. The software does not release these 4 bits automatically after using them to transfer information to the LCD but you do have the option of saving the value of PORTD before using the LCD, and then restoring this value after the LCD has been written to. The complication, of course, being that there will be a short glitch when the LCD is written to and the use you make of PORTD has to tolerate this discontinuity.

PORTE, which has only three external lines, is dedicated to controlling the information transfer to the LCD. These lines can be used for other purposes if the LCD is not being used. PORTE is made digital when controlling the LCD and can be used for analog inputs when its pins are specified as analog inputs. This is done with the ADCON1 register, as described earlier and in Chapter 9.

The LCD provided on the LAB-X1 allows us to display two lines of 20 characters each. Its connections to the microcontroller are shown in Figure 5.3.

Since this is important, let's take another look. In Figure 5.3, we see that the LCD uses all the lines available on PORTD and PORTE. All of PORTD is used as the port the data will be put on, while PORTE, which has only three lines, is used to control data transfer to the LCD. We also know from looking at the full schematics provided with the LAB-X1 that all of PORTD is also connected to the LED bargraph. This does not affect our programming of the LCD and we will ignore this for now. You will, however, see the LEDs in the bargraph flicker ON and OFF as programs run, because we will be manipulating the data on the lines (D0 to D7). It is also possible to feed the LCD with just the 4 high bits of PORTD. See the PBP manual on how to do this. It takes slightly longer to refresh the LCD when you are using only 4 bits, and writing to the LCD is one of the most time-consuming tasks in most programs.

Let's write the ubiquitous "Hello World" program for the LCD as our first exercise in programming the LCD (see Program 5.4). Once we know how to do that, we can basically write whatever we want, when we want, to the LCD display.

54 CONTROLLING THE OUTPUT AND READING THE INPUT

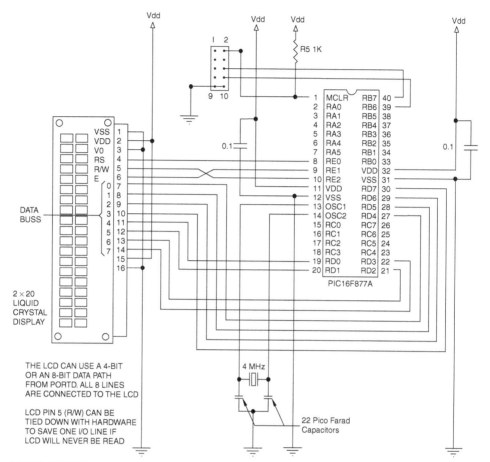

Figure 5.3 The LCD display wiring. (An easy-to-comprehend schematic diagram showing just the lines between the microcontroller and the 2 x 20 display module.)

Program 5.4 Displaying and blinking "HELLO WORLD" in the LCD display

For the LCD display registers on the LAB-X1, the DEFINE statements are as follows:

```
CLEAR                        ; clear the memory
DEFINE OSC   4               ; osc speed              ]
DEFINE LCD_DREG PORTD        ; data register          ]
DEFINE LCD_RSREG PORTE       ; select register        ]
DEFINE LCD_RSBIT 0           ; select bit             ] These DEFINEs
DEFINE LCD_EREG PORTE        ; enable register        ] are all explained
DEFINE LCD_EBIT 1            ; enable bit             ] in the PBP
DEFINE LCD_RWREG PORTE       ; read/write register    ] manual.
```

(Continued)

Program 5.4 Displaying and blinking "HELLO WORLD" in the LCD display (*Continued*)

```
DEFINE LCD_RWBIT 2          ; read/write bit           ]
DEFINE LCD_BITS 8           ; width of data path      ] Can be 4 or 8
DEFINE LCD_LINES 2          ; lines in display        ]
DEFINE LCD_COMMANDUS 2000   ; delay in micro seconds  ]
DEFINE LCD_DATAUS 50        ; delay in micro seconds  ]
                            ;
; Set the port directions. We are setting (must set) all of PORTD
; and all of PORTE as outputs even though PORTE has only three
; lines. (The low nibble in PORTD can be set as inputs if we use
; a 4 high bit path to feed the LCD.)
                            ;
PAUSE 500                   ; allow for LCD startup
TRISD = %00000000           ; set all PORTD lines to output
TRISE = %00000000           ; set all PORTE lines to output
                            ; set the Analog-to-Digital control register
ADCON1=%00000111            ; needed for the 16F877A - see above and
                            ; below - this makes all of ports A and E
                            ; digital.
LOOP:                       ; the main loop of the program
  LCDOUT $FE, 1             ; clear screen, go to position 1
  PAUSE 250                 ; pause 0.25 seconds
  LCDOUT "HELLO"            ; print
  LCDOUT $FE, $C0           ; go to second line, first position
  LCDOUT "WORLD"            ; print
  PAUSE 250                 ; pause 0.25 seconds
GOTO LOOP                   ; repeat
END                         ; all programs must end in END
```

Before we can write to the LCD, we have to define how the LCD is connected to the MCU. *Also, since the 16F877A has some analog capabilities, it will start up and reset in analog mode, and has to be changed to digital mode (for PORTE only in our immediate case) before we can use its digital properties.*

The compiler manual tells us we have to specify the location of the LCD and specify the control lines connected to it so the compiler can address the device properly. Doing so lets us place the LCD wherever it's convenient for us when we design our own devices (and the compiler will still be able find it). These variables are to be specified in DEFINE statements before any of the rest of the program is written. (In this book, we will always place the LCD at the same address locations used by the LAB-X1 so we can test all our programs for our instruments on the LAB-X1 when we need to.)

Program 5.4 demonstrates the most elementary control over output to the LCD display. Variations of these lines of code, and the addition of a few command codes, will be used to write to the LCD in all our programs.

Not all the preceding DEFINE statements are needed on the LAB-X1, but when you build your own devices, you will need to include them all to make sure nothing has been omitted. You never know what state an MCU might start up in, so cover all your bases.

Controlling the Digital and Analog Settings

ADCON1 = **A**nalog to **D**igital **CON**trol register #1.

The ADCON1=%00000111 statement, or one like it, is needed for the 16F877A because any PIC MCU processor that has any analog capabilities at all comes up in the analog mode on reset and startup. This particular instruction puts all the analog pins on ports A and E into the digital mode. Since we need only PORTE and PORTD for controlling the LCD, none of PORTA needs to be in digital mode. I am showing %00000111 because all the examples used by microEngineering Labs in all their literature and on their Web site use and suggest using this value. See datasheet for more detailed information. (The use of this register is also discussed in greater detail in Chapter 9.) If you want to turn just the three available lines on PORTE to digital, you can use any binary value from 010 to 111 inclusive.

The control of the A-to-D conversion capability is managed by the 4 low bits of ADCON1. For our purposes, bit 0 and bit 3 are not relevant. See page 112 of the datasheet for detailed information on this.

Bit 0 is not relevant to the LCD operation (it is a "don't care" bit).

Bit 1 and 2 must be set to 1 to make the two ports (A and E) digital.

Bit 3 is not relevant to the LCD operation (it, too, is a "don't care" bit).

So ADCON1 = %00000110 or %00000111 would be adequate for our work. (We could also have done this in decimal format with ADCON1=6 or with ADCON1=7.)

Writing Binary, Hex, and Decimal Values to the LCD

The value of numbers written to the LCD can be specified with prefixes that determine if the value will be display as a binary, a hexadecimal, or a decimal value. See the PBP manual for details.

BIN specifies that the display will be binary.

HEX specifies that the display will be in hexadecimal format.

DEC specifies that the display will be in decimal format.

In Program 5.5, the value of "NMBR" is set to 170 arbitrarily (actually because it alternates 1s and 0s in binary format). Any number below or equal to 255 could have been used. Using BIN8 instead of BIN displays all 8 bits. Using Hex2 instead of HEX displays both hex digits. DEC5 can display all five decimal digits because we are limited to 16 bits

(65,535) and integer math in PICBASIC. BIN16 can be used for 2-byte words to display all 16 bits. Any number of digits up to 16 can be displayed.

Program 5.5 Writing to the LCD display in FULL binary, hexadecimal, and decimal

```
CLEAR                          ; clear memory
DEFINE OSC 4                   ; osc speed
DEFINE LCD_DREG PORTD          ; define LCD connections
DEFINE LCD_DBIT 4              ; define LCD connections
DEFINE LCD_RSREG PORTE         ; define LCD connections
DEFINE LCD_RSBIT 0             ; define LCD connections
DEFINE LCD_EREG PORTE          ; define LCD connections
DEFINE LCD_EBIT 1              ; define LCD connections
ADCON1=%00000110               ; make PORTA and PORTE digital
LOW PORTE.2                    ; LCD R/W low (write) We will do no reading
PAUSE 500                      ; wait for LCD to start up
                               ;
NMBR VAR BYTE                  ; assign variable
                               ;
TRISD = %00000000              ; D7- -D0 = all outputs
NMBR = %10101010               ; this is decimal 170
                               ;
LCDOUT $FE, 1                  ; clear the LCD
LCDOUT $FE, $80, BIN8 NMBR," ",HEX2 NMBR, " ", DEC5 NMBR," "

                               ; display
END                            ; end program
```

Reading a Potentiometer and Displaying the Results on the LED Bargraph

NOTES ON READING THE POTENTIOMETERS ON THE LAB-X1

Each potentiometer is placed across ground and 5 volts and the wiper is read on the A/D line. (Other reference voltages and resistances can be specified and used—see the datasheet.) The potentiometer value has to be high enough so as not to act as a short between the power and ground. 5K ohms as a minimal value is okay. (Extremely high values make for a jittery reading.)

When we read a potentiometer, the MCU divides the voltage across the potentiometer into 256 steps between 0 and 255 and gives us the number that represents the position of the wiper across the connected voltage. Neither the voltage nor the resistance of the potentiometer is relevant (though it can be if we know the minimum and maximum voltage across the pot). What we are getting is the relative position of the wiper expressed as an 8-bit number. (The PIC also has a 10-bit resolution capability—see the datasheet and PICBASIC manual.)

The overall resistance of each of the three potentiometers is placed across 5 volts, and the three PORTA lines read the position of the three wipers. The potentiometers are read as 8-bit values using an 8-bit A-to-D converter. This gives a full-scale reading of from

0 to 255 for all three potentiometers, no matter what the actual total resistance value of each potentiometer. If you want to read the resistance in ohms, you must divide the reading by 255 and multiply by the total resistance of the potentiometer. (Again, the potentiometer value must be high enough so the potentiometer does not act as a short between ground and the MCU power connection. 5K to 10K ohms is a good selection for most purposes.)

Next, we will read just one of the potentiometers (the one nearest the edge of the board) to an accuracy of 8 bits and display the results on the rightmost eight LEDs of the LED bargraph. This potentiometer is connected to pin 2 of the PIC (also identified as RA0 and as pin PORTA.0). We will display the result of the value read (0 to 255) on the bargraph by loading the reading into PORTD. Since PORTD is connected to the eight LEDs, this will automatically give us a binary representation of the data. In the next step, we will display the information on the LCD display as alphanumeric data (which is, of course, much easier to read).

Expanding the program to not only display to the bargraph but also put the information on the LCD display, we have a problem in that the PORTD lines are shared by the bargraph and the LCD display. When we run the program, we will notice that there is a background noisy blinking of the LEDs in the bargraph as the LCD is written to, but after that is done the bargraph displays the data from the potentiometer as expected. If we had hardware and software that could suppress the LEDs when we were writing to the LCD, this problem could be eliminated. The operation observed demonstrates that the chip select line allows us to use the lines of PORTD to control both the LED bargraph and the LCD display. Once we are done with writing to the LCD we can load the data into PORTD and pause the program to allow us to read the display. Notice that the pause/delay must come immediately after setting PORTD to A2D_Value for this to work properly. When we do what, is important when using microcontrollers. Program 5.6 shows us how this is done and Figure 5.4 shows the relevant circuitry.

Program 5.6 Displaying the potentiometer wiper position on the LCD and the LED bargraph

```
CLEAR                       ; define LCD connections
DEFINE OSC 4                ; osc speed
DEFINE LCD_DREG PORTD       ; define LCD connections
DEFINE LCD_DBIT 4           ; define LCD connections
DEFINE LCD_RSREG PORTE      ; define LCD connections
DEFINE LCD_RSBIT 0          ; define LCD connections
DEFINE LCD_EREG PORTE       ; define LCD connections
DEFINE LCD_EBIT 1           ; define LCD connections
ADCON1=%00000110            ; Make PORTA and PORTE digital
LOW PORTE.2                 ; LCD R/W low (set it to write only)
PAUSE 500                   ; wait for LCD to start up
                            ;
NUMB VAR BYTE               ; assign variable
                            ;
TRISD = %00000000           ; D7 to D0 are all made outputs
A2D_VALUE VAR BYTE          ; create A2D_Value to store result
TRISA = %11111111           ; set PORTA to all input
ADCON1 = %00000010          ; set PORTA analog input
                            ;
```

(Continued)

READING A POTENTIOMETER AND DISPLAYING THE RESULTS ON THE LED BARGRAPH

Program 5.6 Displaying the potentiometer wiper position on the LCD and the LED bargraph (*Continued*)

```
LCDOUT $FE, 1            ; clear the LCD
                         ; define ADCIN parameters
DEFINE ADC_BITS 8        ; set number of bits in result
DEFINE ADC_CLOCK 3       ; set clock source (3=rc)
DEFINE ADC_SAMPLEUS 50   ; set sampling time in uS
                         ;
LOOP:                    ; start loop
  ADCIN 0, A2D_VALUE     ; read channel 0 to A2D_Value
  LCDOUT $FE, $80, "VALUE= ", HEX2  A2D_VALUE, " ", DEC5_
A2D_VALUE
  LCDOUT $FE, $C0, BIN8 A2D_VALUE   ;
  PORTD=A2D_VALUE        ; the pause must come right after setting
                         ; PORTD and before PORTD is used again
  PAUSE 250              ; try setting PORTD before the LCDOUT
GOTO LOOP                ; do it forever
END                      ; end program
```

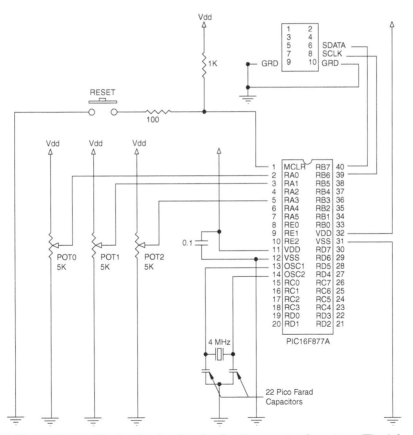

Figure 5.4 The basic circuitry for the three potentiometers. (The information read from potentiometer 0 is displayed on the bargraph in Program 5.6.)

A Simple Beep

We have one other piece of hardware we can output to on the LAB-X1 and that is the small piezoelectric speaker on the board. This speaker is connected to line PORTC.2.

The PWM (Pulse Width Modulation) command can be used to create a short beep on the piezoelectric speaker on the LAB-X1. The command specifies the PORTC pin to be used, the duty cycle and the duration of the beep (100 milliseconds in our case). See Program 5.7.

Program 5.7 Generates a short tone on the piezo speaker (Note that line C2 is HPWM Channel 1)

```
CLEAR                    ; clear memory
DEFINE OSC 4             ; osc speed
PWM PORTC.2, 127, 100    ; beep command
END                      ; end the program
```

Program 5.7 provides a 0.1 second (100 μsec) beep. Circuitry is shown in Figure 5.5. Press the reset button to repeat the beep.

Check to see what happens if you leave the END statement off.

Program 5.7 generates a 50 percent duty cycle for 100 cycles.

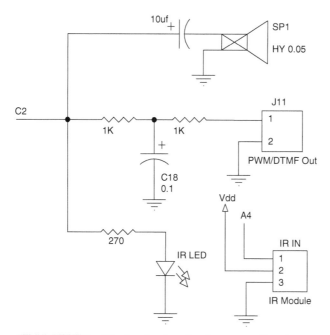

Figure 5.5 The basic circuitry for generating tones on the piezo speaker on the hardware provided. (If you use the infrared receiver, its signal will appear on line A4. Only relevant circuitry is shown.)

It defines that we are using a 4 MHz oscillator.

PORTC.2 specifies the pin to be used.

127 specifies a 50 percent duty cycle; the range of the variable is from 0 to 255.

100 specifies that the tone is to last for 100 of the 256 ON-OFF steps that define one cycle.

In the PWM command, the frequency and length of the signal generated are dependent on the oscillator frequency. In our case, this is 4 MHz, and one cycle is about 2.5 milliseconds long (0.0025 seconds).

Note that the line C2 is also connected to the output for a possible phone jack and to an IR LED that can be used to interact with IR receivers. These two connections are not populated on the PC board as received but they can be added without difficulty.

Two types of signals can be annunciated on the speaker as programmed from the compiler. The PWM command can send a signal of a fixed duty cycle for a fixed number of cycles, and the HPWM command can set up a PWM signal that *runs continuously in the background*. In either case, the signal needs to be provided on the PORTC.2 pin because that is where the speaker is connected. However, the normal PWM (not the background HPWM) command signal can be made to appear at any available pin. The background HPWM signal can be modified "on the fly" in a program.

The HPWM signals can only be made available at pin PORTC.2 (Channel 1) and PORTC.1 (Channel 2). Yes, the pin numbers are reversed! In the PIC 16F877A, there are only two HPWM channels and these two pins are connected permanently to these two channels. (Some PIC devices provide more than two channels. See the datasheets.) Since we have the speaker hardwired to PORTC.2, we can only use HPWM Channel 1 for the tones we generate.

As seen in Figure 5.5, these signals can also be used to generate telephone dial tones (DTMF) and infrared (IR) signals when provided with the appropriate hardware. We will concentrate on creating tones on the piezo speaker. The wiring and programming is the same for the other devices; they are all wired in parallel. Just change the parameters to use them.

Program 5.8 is a slightly more complicated program and demonstrates the use of PWM to control the brightness of one of the LEDs in the bargraph.

Program 5.8 LED dimming using the PWM command

```
CLEAR                         ; clear RAM
DEFINE OSC 4                  ; osc speed
TRISD = %11111110             ; set only PORTD pin 1 to output
X VAR BYTE                    ; declare x as a variable
                              ;
LOOP:                         ; start loop
  FOR X = 0 TO 255 STEP 5   ; ]  in this loop the value
    PWM PORTD.0, X, 3         ; ]  x represents the brightness
  NEXT X                      ; ]  of the LED at PORTD.0
GOTO LOOP                     ; repeat loop
END                           ; end program
```

62 CONTROLLING THE OUTPUT AND READING THE INPUT

Using the HPWM (hardware PWM) command is a bit more complicated in that we must define certain parameters before we can use the command. The necessary defines are as follows:

```
DEFINE CCP1_REG  PORTC    ; Port to be used by HPWM 1
DEFINE CCP1_BIT  2        ; Pin to be used by HPWM 1
DEFINE CCP2_REG  PORTC    ; Port to be used by HPWM 2
DEFINE CCP2_BIT  1        ; Pin to be used by HPWM 2
```

You also have to define which timer the signal will use so other timers can be used for other purposes while the signal is being generated. If a timer is not specified, the system defaults to Timer1, the 16-bit timer.

The command is

```
HPWM Channel, DutyCycle, Frequency
```

The commands

```
DEFINE.OSC 4              ; osc speed
HPWM 1, 127, 1500         ; generate background PWM at 1500 cps.
```

creates a 50 percent duty cycle PWM signal at 1500 Hz (as affected by the definition of OSC) on PORTC.2 continuously in the background. Also see program 5.9.

The command can be updated during runtime from within a program. As might be expected, the pin cannot be used for any other purpose as long as it is generating the PWM signal. Turn OFF the PWM mode at the CCP control register to use the pin as a normal pin. See the datasheet for more information.

The frequencies generated are limited by the frequency of the oscillator being used to clock the PIC processor. The minimum frequency for the PIC 16F877A is 1221 Hz (with a 20-MHz oscillator). Not all frequencies can be generated. See the PICBASIC PRO Compiler manual for more information on other frequencies.

Program 5.9 **Generates a tone on the piezo speaker** (There is no looping in this program)

```
CLEAR                     ; clears memory
DEFINE OSC 4              ; osc speed
DEFINE CCP1_REG PORTC     ; port to be used by HPWM 1
DEFINE CCP1_BIT 2         ; pin to be used by HPWM 1
                          ; since no timer is defined,
                          ; Timer1 will be used;
HPWM 1,127,2500           ; the tone command
PAUSE 100                 ; pause .1 sec to hear tone
END                       ; end program to stop tone.
```

Next, we will generate some telephone touchtone tones on the speaker to demonstrate the capability provided by the DTMFOUT command. (See Program 5.10.)

```
DTMFOUT Pin, {Onms, Offms} [Tone#{Tone#...}]
```

Since we will be using Pin C2, our command will look similar to the preceding because we are using default values for ONms and OFFms

Program 5.10 Generates telephone key tones on the piezo speaker (555-1212)

```
CLEAR                              ; clear memory
DEFINE OSC 4                       ; osc speed
DTMFOUT PORTC.2, [5, 5, 5, 1, 2, 1, 2]   ; telephone tones
END                                ; end program
```

The key tones generated are rough (before filtering), but you can tell that they mimic the telephone dialing tones. The signal needs to go through a filter and then an amplifier to be clean and viable. This command has a number of constraints, depending on the processor used and the speed of the oscillator in the circuit. See the PICBASIC PRO Compiler manual for details.

The FREQOUT command can also be used to generate telephone dialing frequencies. See the PICBASIC PRO Compiler manual for details.

Advanced Exercise: Controlling an RC Servo from the Keyboard

Now that we know how to generate pulses and control the LCD, we can use the LAB-X1 to control the position of an RC servo connected to port J7 from switches SW1, 2, and 3 on the keyboard. Program 5.11 is designed such that . . .

Switch 1 will turn the servo clockwise incrementally.

Switch 2 will center the servo from wherever it is.

Switch 3 will turn the servo counterclockwise incrementally.

The circuitry for this exercise is shown in Figure 5.6.

Note that by changing a few variables that are defined up-front, we can adjust the center position, the incremental step value, and the extreme CW and CCW positions of the servo. (This program has been adapted from, and made simpler than, a program in the microEngineering Labs sample programs. It is instructive to compare this program with the programs SERVOX and SERVO1 in the sample programs.)

Program 5.11 Servo Position Control for an R/C servo from PORTB buttons
(This program uses a servo on Jumper J7)

```
CLEAR                       ; clear memory
DEFINE OSC 4                ; osc speed
DEFINE LCD_DREG PORTD       ; define LCD connections
DEFINE LCD_DBIT 4           ; define LCD connections
DEFINE LCD_RSREG PORTE      ; define LCD connections
DEFINE LCD_RSBIT 0          ; define LCD connections
DEFINE LCD_EREG PORTE       ; define LCD connections
DEFINE LCD_EBIT 1           ; define LCD connections
POS VAR WORD                ; servo position variable
```

(Continued)

Program 5.11 Servo Position Control for an R/C servo from PORTB buttons
(This program uses a servo on Jumper J7) (*Continued*)

```
CENTERPOS VAR WORD              ; servo position variable
MAXPOS.VAR WORD                 ; servo position variable
MINPOS VAR WORD                 ; servo position variable
POSSTEP VAR BYTE                ; servo position step variable
SERVO1 VAR PORTC.1              ; alias servo pin  Use J7 for servo
POS=0                           ; set variables
CENTERPOS =1540                 ; set variables
MAXPOS =2340                    ; set variables
MINPOS =740                     ; set variables
POSSTEP =5                      ; set variables
ADCON1 = %00000111              ; PORTA and PORTE to digital
LOW PORTE.2                     ; LCD R/W low = write
PAUSE 100                       ; wait for LCD to startup
OPTION_REG = $01111111          ; enable PORTB pullups
LOW SERVO1                      ; servo output low
GOSUB CENTER                    ; center servo
LCDOUT $FE, 1                   ; clears screen only
                                ;
MAINLOOP:                       ; main program loop
  PORTB = 0                     ; PORTB lines low to read buttons
  TRISB = $11111110             ; enable first row of buttons on kybd
  IF PORTB.4 = 0 THEN GOSUB LEFT; check if any button is pressed
  IF PORTB.5 = 0 THEN GOSUB CENTER    ; and make a move
  IF PORTB.6 = 0 THEN GOSUB RIGHT     ; accordingly
  LCDOUT $FE, $80, "POSITION = ", DEC4 POS , " " ;
  SERVO1 = 1                    ; start servo pulse
  PAUSEUS POS                   ;
  SERVO1 = 0                    ; end servo pulse
  PAUSE 16                      ; servo update rate about 60 Hz
GOTO MAINLOOP                   ; do it all forever
                                ;
LEFT:                           ; move servo left
  IF POS < MAXPOS THEN POS = POS + POSSTEP ;
RETURN                          ;
                                ;
RIGHT:                          ; move servo right
  IF POS > MINPOS THEN POS = POS - POSSTEP ;
RETURN                          ;
                                ;
CENTER:                         ; center servo
POS = CENTERPOS                 ;
RETURN                          ;
END                             ; end program
```

Now, let's make Program 5.11 more sophisticated by using the three potentiometers on the LAB-X1 board to manipulate the three variables that control the center position, the end positions, and the incremental move of the servo in the Program 5.11. We will use just one variable to adjust both end positions because we have only three potentiometers.

ADVANCED EXERCISE: CONTROLLING AN RC SERVO FROM THE KEYBOARD

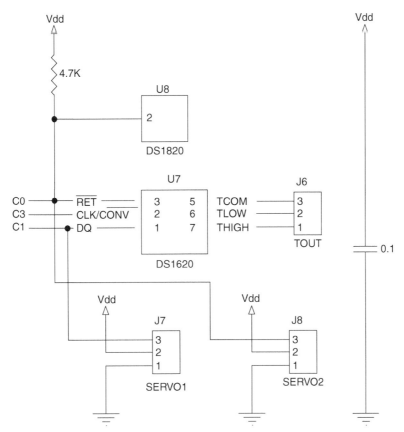

Figure 5.6 Circuitry for controlling an RC servo from the three potentiometers. (As always, only relevant components are shown. Connect the servos to Jumper J7.)

If we had four potentiometers, we could make the adjustment to the limits on one side independent of the adjustment of the other.

We will allow the center position to be adjusted by 127 counts in each direction.

The end positions will be made variable by 127 counts at each end.

The incremental move will be adjustable from 1 to 20 counts per key press.

First, we will make it possible to read the potentiometers. We already know how to do this. Then, we will add the math relationships to the variable in the program so the readings from the potentiometers interact with the three variables appropriately. (See Program 5.12.)

Pot 0, the one nearest the board edge, controls the center position.

Pot 1, the middle pot, controls the limit positions.

Pot 2 sets the speed of the servo by setting the step amount.

Program 5.12 Use servo on jumper pins at the J7 Servo position control, with added functions

```
CLEAR                              ; clear memory
DEFINE  OSC  4                     ; osc speed
DEFINE  LCD_DREG PORTD             ; define LCD connections
DEFINE  LCD_DBIT 4                 ;
DEFINE  LCD_RSREG PORTE            ;
DEFINE  LCD_RSBIT 0                ;
DEFINE  LCD_EREG PORTE             ;
DEFINE  LCD_EBIT 1                 ;
DEFINE  ADC_BITS 8                 ; set number of bits in result
DEFINE  ADC_CLOCK 3                ; set clock source (3=rc)
DEFINE  ADC_SAMPLEUS 50            ; set sampling time in uS
TRISA = %11111111                  ; set PORTA to all input
TRISD = %00000000                  ; set all PORTD lines to outputs
ADCON1 = %00000111                 ; PORTA and PORTE to digital
LOW PORTE.2                        ; LCD R/W line low (W)
A2D_VALUE VAR BYTE                 ; create A2D_Value to store result
A2D_VALUE1 VAR BYTE                ; create A2D_Value to store result
A2D_VALUE2 VAR BYTE                ; create A2D_Value to store result
POS VAR WORD                       ; servo positions
CENTERPOS VAR WORD                 ; center position
MAXPOS VAR WORD                    ; max position
MINPOS VAR WORD                    ; min position
POSSTEP VAR BYTE                   ; step length
PAUSE 500                          ; wait .5 second
SERVO1 VAR PORTC.1                 ; alias servo pin
ADCIN 0, A2D_VALUE                 ; read channel 0 to A2D_Value
OPTION_REG = $7F                   ; enable PORTB pull ups
LOW SERVO1                         ; servo output low
GOSUB CENTER                       ; center servo
LCDOUT $FE, 1                      ; clears screen only
PORTB = 0                          ; PORTB lines low to read buttons
TRISB = %11111110                  ; enable first button row
                                   ;
MAINLOOP:                          ; main program loop
                                   ; check any butn pres'd to move servo
  IF PORTB.4 = 0 THEN GOSUB LEFT     ; handle left move
  IF PORTB.5 = 0 THEN GOSUB CENTER   ; handle centering

  IF PORTB.6 = 0 THEN GOSUB RIGHT    ; handle right move
  ADCIN 0, A2D_VALUE               ; read channel 0 to A2D_Value
  ADCIN 1, A2D_VALUE1              ; read channel 1 to A2D_Value 1
  ADCIN 3, A2D_VALUE2              ; read channel 2 to A2D_Value 2
  MAXPOS =2350 -127 + A2D_VALUE1 ; max position relationship defined
  MINPOS =750 +127-A2D_VALUE1      ; min position relationship defined
  CENTERPOS=POS-127 + A2D_VALUE ; center position relationship defined
  SERVO1 = 1                       ; start servo pulse
    PAUSEUS POS                    ; pulse length
  SERVO1 = 0                       ; end servo pulse
  LCDOUT $FE, $80, "POS=", DEC POS-127 + A2D_VALUE , " ",DEC_
A2D_VALUE," ",DEC A2D_VALUE1," " ,DEC POSSTEP," ";
  PAUSE 16                         ; servo update rate about 60 Hz
```

(Continued)

Program 5.12 Use servo on jumper pins at the J7 Servo position control, with added functions (*Continued*)

```
GOTO MAINLOOP                        ; do it all forever
                                     ;
LEFT:                                ; move servo left
  IF POS < MAXPOS THEN POS = POS + POSSTEP
RETURN                               ;
                                     ;
RIGHT:                               ; move servo right
  IF POS > MINPOS THEN POS = POS - POSSTEP
RETURN                               ;
                                     ;
CENTER:                              ; center servo
  POS = 1540-127 + A2D_VALUE         ;
RETURN                               ;
                                     ;
END                                  ; end program
```

At this stage, we are starting to get an idea about how one might take a simple problem and make it amenable to a more sophisticated solution by adding simple hardware and software features to it. We have gone from a simple but rigid control of the position of a servo to a much more flexible and user-friendly approach. Adding features like these to our instruments and controllers will make them more intuitive, useful, and ergonomic.

Reading the Inputs

Now that we are beginning to learn how to control the output, we need to learn how to read the inputs and manipulate the outputs based on what the input is. In other words, we are going to learn how to create interactive, and thus maybe more useful programs, instruments, and controllers.

READ THE FIRST COLUMN, FIRST ROW PUSH BUTTON (SW1) AND TURN ON AN LED ONLY WHILE THE BUTTON (SW1) IS DOWN

The simplest input is to read just one push button and the simplest output is to turn just one LED ON. We will do just that but we will add a little complication. The LED is to be programmed to be ON only while the button is held down. We will use button 1 (top left) on the keyboard and the LED connected to PORTD.0. This emulates the operation of an ordinary momentary contact switch in any real world application.

READING THE KEYBOARD

On the LAB-X1, all of PORTB is dedicated to the interface with the keyboard. Lines B0 to B3 are connected to the rows, and the lines B4 to B7 are connected to the columns of the keyboard. When the keyboard is not being used, the lines may be used for other purposes, but keep in mind the internal pull-up capability and the in-line load limiting resistors on the lower 4 bits/lines (B0 to B3). These can, of course, easily be made to remain outside our circuitry, so none of this is a problem for us.

In other words, the keyboard is connected to PORTB such that the columns of the keyboard matrix are connected to the high nibble of a port and the rows are connected to the low nibble. The wiring schematic is shown in Figure 5.7.

To read a keyboard like this, the low nibble of PORTB is set to be outputs and the high nibble is set to be inputs.

On the PIC 16F877A, PORTB has a special property that allows its lines to be pulled high (very weakly) with internal resistors by setting OPTION_REG.7 = 0. This property of the PORT can affect all the high bits (B5 to B7) but only those bits that are actually programmed to be inputs with TRISB will be affected.

Next, the four (the low bits, B3 to B0) are made low one line at a time, while the high bits are polled to see if any of them has been pulled low. If any of the switches is down, one of the lines will be pulled low. Because we know which low bit was selected when the high bit became low, we can determine which key has been pressed. For our purposes, at this stage we are interested only in SW1, the upper left switch, so we can simplify the circuitry to what is shown in Figure 5.8 for one row, and then what is shown in Figure 5.9 for just one key.

In these diagrams, we can see that if we make PORTB.0 low and PORTB.4 has been pulled high, PORTB.4 will become low only if SW1 is held down. No polling is necessary at this stage. Once the conditions are set up, all we have to do is create a loop that turns the PORTD.0 LED ON if the switch SW1 is down and OFF for all other conditions. The code for this is listed in Program 5.13.

Program 5.13 Reading a switch (Program reads SW1 and turns LED on PORTD.0 ON while it is down)

```
CLEAR                     ; clear memory
DEFINE OSC 4              ; osc speed
TRISB = %11110000         ; set the PORTB directions
PORTB = %11111110         ; Set only B0 made low.
                          ; See page 31 of the datasheet re:
                          ; the pull ups on PORTB
OPTION_REG.7=0            ; bit 7 of the OPTION_REG sets the pull ups
                          ; when cleared
TRISD = %11111110         ; set only PORTD.0 to an output.
PORTD.0=0                 ; initialize this LED to OFF
                          ;
MAINLOOP:                 ;
  IF PORTB.4=1 THEN       ; check for first column being low
    PORTD.0=0             ; if it is low turn D0 OFF
  ELSE                    ;
    PORTD.0=1             ; if not turn it ON
  ENDIF                   ;
GOTO MAINLOOP             ; repeat.
END                       ;
```

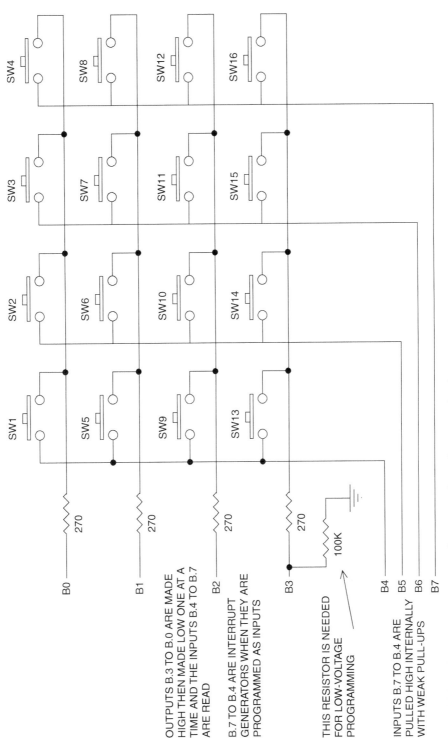

Figure 5.7 Keyboard wiring for the keyboard rows and columns.

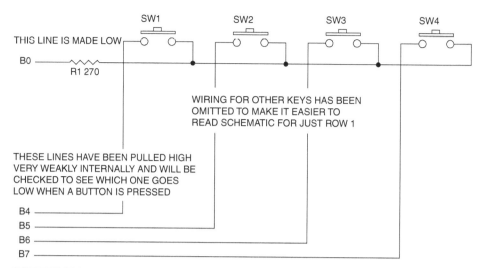

Figure 5.8 Partial keyboard. (The wiring for just one line of switches. The other wiring is still there but is being ignored in the diagram and in the program.)

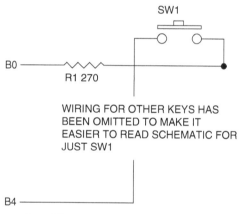

Figure 5.9 Just one key. (The wiring for just one switch. The other wiring is still there but is being ignored in the diagram and in the program.)

Keep in mind that in Program 5.13 we are looking at SW1 only. The other switches in this column will not turn the LED on because they are all high and cannot change the state of PORTB.4 because it is already pulled high (and needs to go LOW if we are to read it as having changed its state).

READ ENTIRE KEYBOARD AND DISPLAY THE BINARY VALUE OF THE ROW AND COLUMN READ ON THE LCD

Next, we learn how to read the entire keyboard and tell which key was pressed by identifying the active row and column numbers. This is a modification of the single key program with the scanning of the nibbles in PORTB added to determine what happened and when it happened.

A loop scans the high nibble of PORTB, the output from the keyboard. When all 4 bits are pulled high, this nibble will be read as HEX F. If it is HEX F, no keys are down and we rescan the keys. If, however, a key has been pressed, the answer will be other than HEX F and can be interpreted as follows:

If B4 is low the answer will be HEX E (15–1=14) 1110 Column 1

If B5 is low the answer will be HEX D (15–2=13) 1101 Column 2

If B6 is low the answer will be HEX B (15–4=11) 1011 Column 3

If B7 is low the answer will be HEX 7 (15–8=7)   0111 Column 4

To determine which row the key that was pressed is in, we have to know which of the bits in the low nibble had been taken LOW by the scanning routine.

The values for the low nibble are as follows:

If B0 is low the low nibble will be HEX E (15–1=14) 1110 Row 1

If B1 is low the low nibble will be HEX D (15–2=13) 1101 Row 2

If B2 is low the low nibble will be HEX B (15–4=11) 1011 Row 3

If B3 is low the low nibble will be HEX 7 (15–8=7)   0111 Row 4

Having the *two pieces of preceding information* lets us identify the key that was pressed. No matter how many keys we have and no matter how they are laid out, the scanning routine to read the keyboard will be something like what was just explained.

Next, in Program 5.14 we will display the contents of the entire byte on the first line of the LCD so we can actually see what is happening in the register represented by PORTB as we scan the lines. Then, on the second line we will show the low byte and the high byte separately so we can see what each key press does. We have added a 1/20 second delay in the loop (so we can see the scanned value), so we have to hold each key down for at least 1/20 second for the scan to make sure the key press will register and show in the display.

Program 5.14   **Read keyboard** (Reading the keyboard rows and columns)

```
CLEAR                           ; clear memory
DEFINE   OSC   4                ; osc speed
DEFINE LCD_DREG PORTD           ; LCD defines follow
DEFINE LCD_DBIT 4               ;
DEFINE LCD_RSREG PORTE          ;
DEFINE LCD_RSBIT 0              ;
DEFINE LCD_EREG PORTE           ;
DEFINE LCD_EBIT 1               ;
ADCON1 = %00000111              ; make PORTA and PORTE digital
```

(*Continued*)

72 CONTROLLING THE OUTPUT AND READING THE INPUT

Program 5.14 **Read keyboard** (Reading the keyboard rows and columns) (*Continued*)

```
LOW PORTE.2                    ; LCD R/W low (write)
PAUSE 500                      ; wait for LCD to start up
                               ;
READING VAR BYTE               ; define the variables
ALPHA VAR BYTE                 ;
BUFFER VAR BYTE                ;
                               ; set up port B pull ups
OPTION_REG.7 = 0               ; enable PORTB pull ups to make B4-B7 high
TRISB = %11110000              ; make B7-B4 inputs, B3-B0 outputs
BUFFER=%11111111               ; no key has been pressed for display
                               ; set up the initial LCD readings
LCDOUT $FE, 1                  ; clear the LCD
LCDOUT $FE, $C0, "ROW=",BIN4 (BUFFER & $0F)," COL=",_
BIN4 BUFFER >>4
                               ;
LOOP:                          ;
  PORTB =%00001110             ; set line B0 low so we can read row 1
  FOR ALPHA = 1 TO 4           ; only need to look at 4 rows
    LCDOUT $FE, $80, BIN8 PORTB," SCANVIEW B" ; see bits scanned
    IF (PORTB & $F0)$F0 THEN   ; as soon as one of the bits in B4
                               ; to B7 changes we immediately
                               ; have to store the value of PORTB
      BUFFER =PORTB            ; in a safe place.
      GOSUB SHOWKEYPRESS       ;
    ELSE                       ;
    ENDIF                      ;
      PAUSE 50                 ; pause lets us see the scan but
                               ; it also means hold a key down
                               ; for over 50 usecs to have it
                               ; register. Pause be removed
                               ; after you have seen the bits
                               ; scanning on the LCD
    PORTB= PORTB <<1           ; move bits left one place for next
                               ; line low
    PORTB= PORTB + 1           ; put 1 back in LSBit, the right bit
  NEXT ALPHA                   ;
GOTO LOOP                      ;
                               ;
SHOWKEYPRESS:                  ;
  LCDOUT $FE, $C0, "ROW=", BIN4  (BUFFER & $0F),_
" COL=", BIN4   BUFFER >>4     ;
RETURN                         ;
END                            ;
```

Read Keyboard and Display Key Number on the LCD

Now that we understand how this works, we have to turn the binary information we have gathered into a number from 1 to 16 and identify the key press on the LCD. (The sample program to do this, which is provided on the Internet by microEngineering Labs, shows another way of doing this and is worth studying.)

The switch number is the row number plus (the column number −1) * 4.

If we reverse all the bits in the PORTB byte, the nibbles will give us the positions of the rows and columns as the locations of the 1s in the two nibbles. Make sure you understand this before proceeding. Work it out on a piece of paper step by step.

The "Show Key Press" program must be modified to appear as shown in Program 5.15.

Program 5.15 **Reading the keyboard** (Reading the keyboard rows and columns and show key number)

```
CLEAR                        ; clear memory
DEFINE OSC   4               ; osc speed
DEFINE LCD_DREG  PORTD       ; define LCD connections
DEFINE LCD_DBIT  4           ;
DEFINE LCD_RSREG PORTE       ;
DEFINE LCD_RSBIT 0           ;
DEFINE LCD_EREG  PORTE       ;
DEFINE LCD_EBIT  1           ;
ADCON1 = 7                   ; make PORTA and PORTE digital
LOW PORTE.2                  ; LCD R/W low (write)
PAUSE 200                    ; wait for LCD to start
                             ; define the variables
BUFFER VAR BYTE              ;
ALPHA VAR BYTE               ; counter for rows
COLUMN VAR BYTE              ;
ROW VAR BYTE                 ;
SWITCH    BYTE               ;
                             ; set up PORTB pullups
OPTION_REG.7 = 0             ; enable PORTB pullups to make B4-B7 high
TRISB = %11110000            ; make B7-B4 inputs, B3-B0 outputs
                             ; set up the initial LCD readings
LCDOUT $FE, 1                ; clear the LCD
LOOP:                        ;
  PORTB =%00001110           ; set line B0 low so we can read row 1 only
```

(*Continued*)

Program 5.15 **Reading the keyboard** (Reading the keyboard rows and columns and show key number) (*Continued*)

```
    FOR ALPHA = 1 TO 4            ; need to look at 4 rows
      IF (PORTB & $F0)<>$F0 THEN  ;
                                  ; as soon as one of the bits in
                                  ; B4 to B7 changes we; immediately
                                  ; have to store the value of PORTB
        BUFFER =PORTB             ;
        GOSUB SHOWKEYPRESS        ;
      ELSE                        ;
      ENDIF                       ;
      PORTB= PORTB << 1           ; move bits left one place for
                                  ; next line low
      PORTB= PORTB + 1            ; put 1 back in LSBit, the right bit
    NEXT ALPHA                    ;
GOTO LOOP                         ;
                                  ;
SHOWKEYPRESS:                     ;
    BUFFER = BUFFER ^ %11111111   ; flips all the bits in the buffer
                                  ; print the first line
    LCDOUT $FE, $80, "ROW=",BIN4 (BUFFER & $0F),"  _
COL=", BIN4 BUFFER >>4            ;
    COLUMN=(NCD BUFFER) -4        ; calculate column
    ROW=NCD (BUFFER &$0F)         ; calculate row
    SWITCH=((ROW-1) * 4) +COLUMN  ; calculate switch number
                                  ; print the second line
    LCDOUT $FE, $C0, "ROW=", DEC ROW, " COL=", DEC COLUMN, _
" SW=", DEC SWITCH, " "           ;
RETURN                            ;
END                               ;
```

Read One Potentiometer and Display Its 8-Bit Value on LCD in Binary, Hex, and Decimal Notation, also Impress the Binary Value on the Bargraph

A detailed discussion of A-to-D conversions is covered in the chapter, which is devoted to the construction of a digital thermometer instrument based on this capability.

As mentioned before, the potentiometers are read by dividing the voltage across a potentiometer into 256 parts and seeing which of the 256 divisions match the position of the wiper. This gives a reading between 0 and 255 (in 8-bit resolution). It does not tell us anything about the resistance of the potentiometer, only the relative position of the wiper.

We will read/use the pot closest to the edge of the board. This pot is connected to line PORTA.0, which is pin2 of the MCU.

A-to-D conversions are controlled by the **ADCON0** and **ADCON1** registers, and the 16F877A has to be in **analog mode for the relative pins** for A-to-D conversions to be possible.

Setting the bits in **ADCON0**. (See page 111 of the datasheet.)

Bits 7 and 6 control the clock/oscillator to be used. Set these both to 1.

Bits 5 to 3 select which channels are to be used in the conversions; set to 000 for PORTA.0.

Bit 2 cleared when the conversion is completed. Set it to 1 to start the conversion.

Bit 1 ignored in A/D conversions. Set it to 0.

Bit 0 controls A-to-D conversions. Set it to 1 to enable A-to-D conversions.

When the conversion is completed, the result will be placed in ADRESH and ADRESL. The format of how this is done depends on how the result is set up with register ADCON1.

ADCON1 needs bit 7 to be set to 0 to make the 8-bit result appear in ADRESH and bit 2 needs to be set to 1 to select potentiometer 0 and set the proper reference voltages. See page 112 of the datasheet.

So we set **ADCON0** to %11000001 to set up for reading PORTA.0.

And we set **ADCON1** to %00000010.

The *program segment* to read a value is shown in Program 5.16.

Program 5.16 **Potentiometer readings** (Displaying the value of the potentiometer in all formats)

```
DEFINE OSC   4                    ; osc speed
LOOP:                             ; begin loop
  ADCON0.2 = 1                    ; start conversion
  NOT_DONE:                       ; marker if not done
    PAUSE 5                       ;
  IF ADCON0.2 = 1 THEN NOT_DONE   ; wait for low on bit-2 of ADCON0, conv
  A2D_VALUE = ADRESH              ; move high byte of result to A2D_Value
  LCDOUT $FE, 1                   ; clear screen
    LCDOUT "VALUE: ", DEC A2D_VALUE," "   ; display the decimal value
  PAUSE 100                       ; wait 0.1 second
GOTO LOOP                         ; do it forever
```

(*Continued*)

Program 5.16 **Potentiometer readings** (Displaying the value of the potentiometer in all formats) (*Continued*)

The *complete program* would look like the following:

```
DEFINE LCD_DREG PORTD           ; define LCD registers and bits
DEFINE LCD_DBIT 4               ;
DEFINE LCD_RSREG PORTE          ;
DEFINE LCD_RSBIT 0              ;
DEFINE LCD_EREG PORTE           ;
DEFINE LCD_EBIT 1               ;
A2D_VALUE VAR BYTE              ; create A2D_Value to store result
                                ; set PORTA set PORTD
TRISA = %11111111               ; wet PORTA to all input
TRISD = %00000000               ; wet PORTD to all output
ADCON0 = %11000001              ; configure and turn on A/D Module
ADCON1 = %00000010              ; set PORTA analog and LEFT justify
PAUSE 500                       ; wait 0.5 second for LCD  startup
                                ;
LOOP:                           ;
  ADCON0.2 = 1                  ; start conversion
  NOT_DONE:                     ;
  IF ADCON0.2 = 1 THEN NOT_DONE ; wait for low on bit-2 of ADCON0,
                                ; conversion finishes
  A2D_VALUE = ADRESH            ; move high byte of result to A2D_Value
  LCDOUT $FE, 1                 ; clear screen
  LCDOUT "DEC VALUE= ", DEC A2D_VALUE," "   ; Display 3 values
  LCDOUT $FE, $C0, "HEX=", HEX2 A2D_VALUE," ","BIN=", BIN8_
A2D_VALUE," "                   ;
  PORTD=A2D_VALUE               ; displays value in bargraph
  PAUSE 100                     ; wait 0.1 second
GOTO LOOP                       ; do it forever
END                             ; end program
```

In Program 5.16, we used the named registers themselves to set up the conversions. In the program in the next section, we will use the power of the compiler and its related commands to read the three pots much more conveniently by using the ADCIN command.

Read All Three Potentiometers and Display Their Values on the LCD

Five of the six pins on PORTA can be used as analog inputs. In our case, pins 0, 1, and 3 are connected to the three potentiometers. (Pin A4 cannot be used.)

READ ALL THREE POTENTIOMETERS AND DISPLAY THEIR VALUES ON THE LCD

If we want to read all three pots, we have to activate their three lines and create variables to store the three results obtained. The modifications to Program 5.16 are shown in Program 5.17.

Program 5.17 **Display potentiometer settings** (Reading and displaying all three potentiometers values in decimal format)

```
CLEAR                           ; define LCD connections
DEFINE OSC  4                   ; osc speed
DEFINE LCD_DREG PORTD           ;
DEFINE LCD_DBIT 4               ;
DEFINE LCD_RSREG PORTE          ;
DEFINE LCD_RSBIT 0              ;
DEFINE LCD_EREG PORTE           ;
DEFINE LCD_EBIT 1               ;
LOW PORTE.2                     ; LCD R/W line low (W)
PAUSE 500                       ; wait .5 second for LCD startup
                                ; the next 3 defines are needed for
                                ; the ADCIN command
DEFINE ADC_BITS 8               ; set number of bits in result
DEFINE ADC_CLOCK 3              ; set internal clock source (3=rc)
DEFINE ADC_SAMPLEUS 50          ; set sampling time in uS
                                ;
TRISA = %11111111               ; set PORTA to all input
TRISD = %00000000               ; set all PORTD lines to outputs
ADCON1 = %00000110              ; PORTA and PORTE to digital
A2D_Value0 VAR BYTE             ; create A2D_Value to store result 1
A2D_Value1 VAR BYTE             ; create A2D_Value to store result 2
A2D_Value2 VAR BYTE             ; create A2D_Value to store result 3
                                ;
LCDOUT $FE, 1                   ; clear the display
                                ;
MAINLOOP:                       ; main program loop
                                ; check potentiometer values
  ADCIN 0, A2D_VALUE0           ; read channel 0 to A2D_Value0
  ADCIN 1, A2D_VALUE1           ; read channel 1 to A2D_Value1
  ADCIN 3, A2D_VALUE2           ; read channel 2 to A2D_Value2
  LCDOUT $FE, $80, DEC A2D_VALUE0," ",DEC A2D_VALUE1," " ,DEC_
A2D_VALUE2," "                  ;
  PAUSE 10                      ;
GOTO MAINLOOP                   ; do it all forever
END                             ; end program
```

Adding the Kind of Flexibility That Defines Computer Interfaces and Demonstrates the Ability to Make Sophisticated Real-Time Adjustments

Program 5.18 is *similar to Program 5.16 that was developed earlier* but shows another approach.

Use the three potentiometers to control one R/C servo.

Control the relative location of the center position with POT0.

Control limit position of the end positions with POT1.

Control the rate of movement with POT2.

Program 5.18 Servo/Potentiometers (Three potentiometers controlling one servo; connect the servo to Jumper J7 for this program)

```
CLEAR                           ;
DEFINE OSC 4                    ; osc speed
DEFINE LCD_DREG PORTD           ; define LCD connections
DEFINE LCD_DBIT 4               ;
DEFINE LCD_RSREG PORTE          ;
DEFINE LCD_RSBIT 0              ;
DEFINE LCD_EREG PORTE           ;
DEFINE LCD_EBIT 1               ;
LOW PORTE.2                     ; LCD R/W line low (W)
DEFINE ADC_BITS 8               ; set number of bits in result
DEFINE ADC_CLOCK 3              ; set clock source (3=rc)
DEFINE ADC_SAMPLEUS 50          ; set sampling time in uS
TRISA = %11111111               ; set PORTA to all input
TRISD = %00000000               ; set all PORTD lines to outputs
ADCON1 = %00000111              ; PORTA and PORTE to digital
A2D_VALUE VAR BYTE              ; create A2D_Value to store result
A2D_VALUE1 VAR BYTE             ; create A2D_Value1 to store result
A2D_VALUE2 VAR BYTE             ; create A2D_Value2 to store result
POS VAR WORD                    ; servo positions
CENTERPOS VAR WORD              ;
MAXPOS VAR WORD                 ;
MINPOS VAR WORD                 ;
POSSTEP VAR BYTE                ;
PAUSE 500                       ; wait .5 second
```

(*Continued*)

Program 5.18 **Servo/Potentiometers** (Three potentiometers controlling one servo; connect the servo to Jumper J7 for this program) (*Continued*)

```
SERVO1 VAR PORTC.1              ; alias servo pin
ADCIN 0, A2D_VALUE              ; read channel 0 to A2D_Value
OPTION_REG = $01111111          ; enable PORTB pullups
LOW SERVO1                      ; servo output low
GOSUB CENTER                    ; center servo
LCDOUT $FE, 1                   ; clears screen only
PORTB = 0                       ; PORTB lines low to read buttons
TRISB = %11111110               ; enable first button row
                                ; main program loop
MAINLOOP:                       ; check any but. pressed to move servo
  IF PORTB.4 = 0 THEN GOSUB LEFT   ;
  IF PORTB.5 = 0 THEN GOSUB CENTER ;
  IF PORTB.6 = 0 THEN GOSUB RIGHT  ;
  ADCIN 0, A2D_VALUE            ; read channel 0 to A2D_Value
  ADCIN 1, A2D_VALUE1           ; read channel 1 to A2D_Value 1
  ADCIN 3, A2D_VALUE2           ; read channel 2 to A2D_Value 2
  MAXPOS =1500 + A2D_VALUE1*3   ;
  MINPOS =1500 - A2D_VALUE1*3   ;
  CENTERPOS=1500+3*(A2D_VALUE-127)  ;
  POSSTEP =A2D_VALUE2/10 +1     ;
  SERVO1 = 1                    ; start servo pulse
  PAUSEUS POS                   ;
  SERVO1 = 0                    ; end servo pulse
  LCDOUT $FE, $80, "POS=", DEC POS , " "` ;
  LCDOUT $FE, $C0, DEC A2D_VALUE," ",DEC A2D_VALUE1," " , DEC_
POSSTEP," "
  PAUSE 10                      ; servo update rate about 60 Hz
GOTO MAINLOOP                   ; do it all forever
                                ; move servo left
LEFT: IF POS < MAXPOS THEN POS = POS + POSSTEP
RETURN                          ;
                                ; Move servo right
RIGHT: IF POS > MINPOS THEN POS = POS - POSSTEP
RETURN                          ;
                                ; center servo
CENTER: POS = 1500+3*(A2D_VALUE-127)     ;
RETURN                          ;
END                             ; end program
```

Exercises

Caution: Thinking required!

Answers to these problems *are not provided*. There are no unique solutions.

Since making instruments and controllers is really all about inputs and outputs and what you do with them, a comprehensive set of exercises that focus specifically on inputs and outputs are provided.

LED EXERCISES: CONTROLLING THE LIGHT EMITTING DIODES (LEDs)

We will learn about controlling the output from a PIC by writing a series of increasingly complicated programs that will control the ten-segment LED bargraph provided on the LAB-X1. In these exercises, we are controlling the LEDs, but the control strategies developed will apply to any kind of "ON OFF" devices we have connected to the LAB-X1 or to any other device we design.

1. Light the eight LEDs on the right one at a time till they are all lit, and then turn them OFF one at a time. Time delay between actions is to be one-tenth of a second (exactly).
2. Modify the preceding program so the delay time is controlled by the top most potentiometer on the LAB-X1. The time is to vary from 10 milliseconds to 200 milliseconds inclusive, no less, no more.
3. Write a program that will vary the glow on the rightmost LED from fully OFF to fully ON once a second. Program the second LED to go dark and bright exactly 180 degrees out of phase with the first LED so that as one LED is getting brighter, the other LED is getting dimmer and vice versa.
4. Write a program that flashes the four leftmost LEDs ON and OFF every 0.25 seconds and cycles the four LEDs on the right through a bright/dim cycle every two seconds.
5. Write a program that flashes the first LED ten times a second, flashes the second one nine times a second, and flashes the third LED whenever both LEDs are on at the same time. Display how many times the third LED has blinked on the LCD display. (Timing can be approximate but has to have a common divider so the third LED will give the beat frequency.)

LIQUID CRYSTAL DISPLAY EXERCISES: CONTROLLING THE LIQUID CRYSTAL DISPLAY (LCD)

The addresses of the memory locations used by the LCD have already been fixed, as has the instruction set we use to write to the LCD. The description of the Hitachi HD44780U (LCD-II) controller instruction set as well as its electronic characteristics are provided in the data file for the display. Here, we will list only the codes that apply to our immediate use of the device.

Two types of commands can be sent to the display: the control codes and the set of actual characters to be displayed. Both uppercase and lowercase characters are supported, as are a number of special and graphic characters. The control codes allow you to control the display and set the position of the cursor, and soon. Each control code must be preceded by decimal 254 or Hex $FE. (The controller also supports the display of a set of Japanese characters that are not of interest to us.)

Command codes for the following actions are provided along with others. Go to the datasheet for the controller to find out what all these command codes are.

Clear the LCD

Return home

Go to beginning of line 1

Go to beginning of line 2

Go to a specific position on line 1

Go to a specific position on line 2

Show the cursor

Hide the cursor

Use an underline cursor

Turn on cursor blink

Move cursor right one position

Move cursor left one position

There are still other commands to discover in the datasheet. There are memory locations within the LCD, as well as invisible locations beyond the end of the visible 20 characters. You should know how to find all this information.

It is also possible to design your own font for use with this particular display. All the information needed to do so can be found in the Hitachi HD44780U manual/datasheet.

1. Write a program to put the 26 letters of the alphabet and the ten numerals in the 40 spaces that are available on the display. Put four spaces between the numbers and the alphabet to fill in the four remaining spaces. Once all the characters have been entered, scroll the 40 characters back and forth endlessly though the two lines of the display.
2. Write a program to bubble the 26 capital letters of the alphabet through the numbers 0 to 9 on line two of the LCD. (This means: First put the numbers on line two. "A" then takes the place of the "0" and all the numbers move over. Then the "A" takes the place of the "1" and the "0" moves to position 1. Afterward, the "A" replaces the "2" and so on till it gets past the 9. Following this, the "B" starts its way across the numbers and so on.) Loop forever.
3. Write a program to write the numbers 0 to 9 upside down on line 1. Wait 1 second and then flip the numbers right side up. Loop.
4. Create a program to write "HELLO WORLD" to the display and then change it to lowercase one letter at a time with 100 milliseconds between letters. Wait 1 second and go back to uppercase one character at a time with negative letters (all dots on the display are reversed to show a dark background with white letters in lowercase). Loop.

MISCELLANEOUS EXERCISES

These exercises are designed to challenge your programming ability. Again, you will need access to the datasheet for the LC Display.

1. Editor: Write a program that displays a random 12 numbers on line 1 of the LCD and displays a cursor that can be moved back and forth across the 20 spaces with potentiometer 0. The entire range of the potentiometer must be used to move across the 20 spaces. Allow the keypad to insert numbers 0 to 9 into the position the cursor is on. Assign a delete switch and an insert space switch on the keyboard. A comprehensive number (plus decimal and space) editor is required.

2. Mirror: Write a program that puts a random set of letters and numbers on line 1 and then puts their mirror images on line 2. The mirror is between line 1 and line 2. You have to learn how to create the upside-down numbers from the Hitachi datasheet for the display, and also learn how to read what is in the display from the display ROM.

3. Forty Characters: The display ROM is capable of storing 40 characters on each line. Design a program to allow you to scroll back and forth to see all 40 characters on both lines one line at a time. Use two potentiometers for scrolling, one for each line.

4. Four lines: Write a program to display four lines of random data on the LCD and to scroll up and down and side to side to see all four lines in their entirety. You have to store what is lost from the screen before it is lost so you can re-create it when you need it.

5. Bargraphs: Create a three-bargraph display, with each bar 3 pixels high, that extends across both lines of the LCD. The lengths of the bargraphs are determined by the settings of the three potentiometers, which change as the potentiometers are manipulated.

By now, you should be getting pretty good at using the 16F877A and are nearly ready to finish the introduction. Only a little more and we will be ready for just that!

6

TIMERS AND COUNTERS

General

If you have no knowledge about timers, you should read this chapter carefully before taking on the chapters that follow. Those chapters provide a much more advanced discussion of the devices based on their usage, as opposed to the introductory approach provided here. However, there is some repetition, so as to allow each part of this book to stand independently.

Most users will find that using timers and counters is the hardest part of learning how to use PIC microcontrollers. With this in mind, we will proceed in a step-by-step manner and build up the programs in pieces that are easier to digest. Once you get comfortable with their setup procedures, you will find that timers and counters are not so intimidating.

We will cover timers and counters separately. Counters are essentially timers that get their clock input from an outside source. There are two counters in the 16F877A, and they are associated with Timer0 and Timer1. Timer2 cannot be used as a counter because it has no way to read an external signal into this timer.

Note *The clock frequency utilized by the timers is one-fourth of the oscillator frequency. This is the frequency of the instruction clock. This means that the counters are affected by every fourth count of the main oscillator. The frequency is referred to as Fosc/4 in the literature. When responding to an external clock signal, the response is to the actual frequency of the input.*

Caution *The PICBASIC PRO Compiler generates code that does not respond to interrupts while a compiled instruction is being executed. Therefore, long PAUSEs (meaning long enough to lose an interrupt signal, depending on how the timer is set up) can lead to lost interrupts if more than one interrupt occurs during the pause. Since interrupts are used for the express purpose of handling critical response/timing needs, this is most undesirable. PAUSE commands should be used with care under these conditions. The program samples provided next give examples of how they can be written to generate shorter pauses.*

Timers

Timer0 will be covered in more detail as a prototypical timer, and discussion and examples for the use of Timer1 and Timer2 will be provided.

The use of timers internal to microprocessors is a bit more complicated than what we have been doing so far because there is a considerable amount of setup required before the timer can be used, and also because the options for setting up the timers are extensive. We will cover the timers one at a time in an introductory manner, but be warned that there is an entire manual (*PICmicro Mid-Range MCU Family Reference Manual* [DS33023]) available from Microchip Technology Inc. that covers nothing but timers, so our coverage here will, of necessity, be rudimentary.

Understanding timers has to do with understanding how to turn them ON and OFF and how to read and set the various bits and bytes that relate to them. Essentially, in the typical timer application, you turn a timer ON by turning on its enable bit. The timer then counts a certain number of clock cycles and sets an interrupt bit, thus causing an interrupt. Your program responds to the interrupt by executing *a specific interrupt handling routine* and then clearing the interrupt bit. The program then returns to wherever it was when the interrupt occurred. The pre/postscalars have to do with modifying the time it takes for an interrupt to take place. The hard part is finding out which bit does what and where it is located, so reading and understanding the datasheet chapter on the timer you are using is imperative. There is no escaping this horror!

Timers allow the microcontroller to create and react to chronological events. These include:

- Timing events
- The creation of clocks for various purposes
- Generating timed interrupts
- Controlling PWM generation
- Waking the PIC from its sleep mode at intervals to do work (and back to sleep)
- Special use of the watchdog timer

The PIC 16F877A has three internal timers. There is also the watchdog timer, which is discussed after the standard timers.

- **Timer0** An 8-bit free-running timer/counter with an optional prescalar. It's the simplest of the timers to use.
- **Timer1** A 16-bit timer that can be used as a timer or as a counter. It is the only 16-bit timer that can be used as a counter. It is also the most complicated of the timers.
- **Timer2** An 8-bit timer with a prescalar and postscalar and cannot be used as a counter. There is no input line for this timer.

Each timer has a timer control register that sets the options and properties that the timer will exhibit. All the timers are similar and each of them has special features that give it special properties. It is imperative you refer to your datasheet for the PIC 16F877A as you experiment with the timer functions. Once you start to understand what the PIC designers are up to with the timer functions, it will start to come together in your mind.

Timers can have prescalars and/or postscalars associated with them that can be used to multiply the timer setting by a limited number of integer counts. The scaling ability is not adequate to allow all exact time intervals to be created but is adequate for all practical purposes. To the inability to create perfectly timed interrupts, we have to add the uncertainty in the frequency of the oscillator crystal, which is usually not exactly what it is stated to be (and which is affected by the ambient temperature as the circuitry warms up). Though fairly accurate timings can be achieved with the hardware as received, additional software adjustments may have to be added if extremely accurate results are desired. The software can be designed to make a correction to the timing every so often to make it more accurate. We will also need an external source that is at least as accurate as we want our timer to be, so we can verify the accuracy of the device we create.

Timer0

Let's write a simple program to see how Timer0 works. We will use the LED bargraph to show us what is going on inside the microcontroller.

As always, the bargraph is connected to the eight lines of PORTD of the LAB-X1.

First, let's write a program that will light the two LEDs connected to D0 and D1 alternately. (See Program 6.1.) Having them light alternately lets you know that the program is running, or more accurately, it lets you know that the segment of the program that contains this part of the code is working. These two LEDs will be used to represent the foreground task in our program. There is no timer process in this program at this stage. There are no interrupts. The program just blinks the LEDs.

Program 6.1 Foreground program blinks two LEDs alternately (No timer is being used in this program at this time)

```
CLEAR                       ; clears all memory locations
DEFINE OSC 4                ; using a 4-MHz oscillator here
TRISD = %11110000           ; make D0 to D3 outputs
PORTD.0 = 0                 ; turn off bit D0
PORTD.1 = 1                 ; turn on bit D1
ALPHA VAR WORD              ; Set up a variable for counting
                            ;
MAINLOOP:                   ; main loop
  IF PORTD.1 = 0 THEN       ; the next lines of code turn the LEDs ON
    PORTD.1 = 1             ; if they are OFF
    PORTD.0 = 0             ;
  ELSE                      ;
    PORTD.1 = 0             ; and OFF if they
    PORTD.0 = 1             ; are ON
  ENDIF                     ;
  FOR ALPHA = 1 TO 300      ; this loop replaces a long pause command
    PAUSEUS 100             ; with short pauses that are essentially
  NEXT ALPHA                ; independent of the clock frequency.
GOTO MAINLOOP               ; do it all forever
END                         ; all programs need to end with END
```

The use of the PAUSEUS loop in the Program 6.1 provides a latency of 100 microseconds (worst case) in the response to an interrupt and eliminates most of the effect of changing the OSC frequency if that should become necessary. It is better than using an empty counter, which would be completely dependent of the frequency of the system oscillator. (There is an assumption here that the 100 μsec latency is completely tolerable to the task at hand, and it is for this program. It may not be for your real-world program though, so it may need to be adjusted.)

We are turning one LED OFF and another LED ON to provide a more positive feedback. As long as we are executing the main loop, the LCDs will light alternately and provide a dynamic feedback of the operation of the program in the foreground loop.

We select a relatively fast ON-OFF cycle so we will better be able to see minor delays and glitches that may appear in the operation of the program as we proceed.

Run this program to get familiar with the operation of the two LEDs. Adjust the counter (the 300 value) to suit your taste.

Next, we want to add the code that will interrupt this program periodically and make a third LED go ON and then OFF using an approximately one-second cycle. This will serve as the interrupt-driven task we are interested in learning how to create. This is the IMPORTANT task in this particular exercise.

Here is what must be added to the program to get the interrupt-driven LED operational.

Enable Timer0 and its interrupts with appropriate register/bit settings.

Add the ON INTERRUPT command to tell the program where to go when an interrupt occurs.

Set up the interrupt routine to do what needs to be done.

The interrupt routine counts to 61 and turns the LED ON if it is OFF and OFF if it is ON.

Clear the interrupt flag that was set by Timer0.

Send the program back to where it was interrupted with the RESUME command.

WHY ARE WE USING 61?

The prescalar is set to 64 (bits 0 to 2 are set at 101 in the OPTION_REG).

The counter interrupts every 256 counts.

$256 \times 64 = 16,384$

Clock is at 4,000,000 Hz

Fosc/4 is 1,000,000

$1,000,000 / 16,384 = 61.0532$. Its not exactly 61, but it is close enough for our purposes for now.

The lines of code now look like those in Program 6.2.

Program 6.2 Using TIMER0 (Program blinks two LEDs [D1 and D0] alternately and blinks a third LED [D2] for one second ON and one second OFF as controlled by the interrupt signal)

```
CLEAR                        ; clear memory
DEFINE OSC 4                 ; using a 4-MHz oscillator
                             ;
OPTION_REG=%10000101         ; page 48 of datasheet
                             ; bit 7=1 disable pull ups on PORTB
                             ; bit 5=0 selects timer mode
                             ; bit 2=1 }
                             ; bit 1=0 } sets Timer0 prescalar to 64
                             ; bit 0=1 }
                             ;
INTCON=%10100000             ; bit 7=1 Enables all unmasked interrupts
                             ; bit 5=1 Enables Timer0 overflow interrupt
                             ; bit 2 flag will be set on interrupt and
                             ; has to be cleared in the interrupt
                             ; routine. It is set clear at start
ALPHA VAR WORD               ; this variable counts in the PauseUS loop
BETA VAR BYTE                ; this variable counts the 61 interrupt
                             ; ticks
TRISD = %11110100            ; sets the 3 output pins in the D port
PORTD = %00000000            ; sets all pins low in the D port
BETA=0                       ;
ON INTERRUPT GOTO INTROUTINE ; this line needs to be early in
                             ; the program,
                             ; in any case, before the routine is called.
                             ;
MAINLOOP:                    ; main loop blinks D0 and D1 alternately
  IF PORTD.1 = 0 THEN        ; ]
    PORTD.1 = 1              ; ]
    PORTD.0 = 0              ; ] this part of the program blinks two
                             ; LEDs in
  ELSE                       ; ] the foreground as described before
    PORTD.1 = 0              ; ]
    PORTD.0 = 1              ; ]
  ENDIF                      ; ]
                             ;
  FOR ALPHA = 1 TO 300       ; the long pause is eliminated with this
                             ; loop
    PAUSEUS 100              ; PAUSE command with short latency
  NEXT ALPHA                 ;
GOTO MAINLOOP                ; end of loop
                             ;
DISABLE                      ; DISABLE//ENABLE must brkt the
                             ; interrupt routine
```

(Continued)

Program 6.2 Using TIMER0 (Program blinks two LEDs [D1 and D0] alternately and blinks a third LED [D2] for one second ON and one second OFF as controlled by the interrupt signal) (*Continued*)

```
INTROUTINE:                  ; this information is used by the compiler
                             ; only.
  BETA = BETA + 1            ;
  IF BETA < 61 THEN ENDINTERRUPT  ; one second has not yet passed
  BETA = 0 ;                 ;
  IF PORTD.3 = 1 THEN        ; interrupt loop turns D3 on and off every
    PORTD.3 = 0              ; 61 times through the interrupt routine.
  ELSE                       ; That is about one second per full cycle
    PORTD.3 = 1              ;
  ENDIF                      ;
ENDINTERRUPT:                ;
  INTCON.2 = 0               ; clears the interrupt flag.
RESUME                       ; resume the main program
ENABLE                       ; DISABLE and ENABLE must bracket
                             ; the interrupt routine
END                          ; end program
```

Make your predictions and then...

Try changing the 3 low bits in OPTION_REG to see how they affect the operation of the interrupt.

In Program 6.2, Timer0 is running free and providing an interrupt every time its 8-bit counter overflows from FF to 00. The prescalar is set to 64 so we get the interrupt after 64 of these cycles. When this happens, we jump to the "IntRoutine" routine, where we make sure that 61 interrupts have taken place, and if they have, we change the state of an LED and return to the place where the interrupt took place. (It happens that it takes approximately 61 interrupts to equal one second in this routine with a processor running at 4 MHz. This could be refined by trial and error after the initial calculation.)

Note that the interrupt is disabled while we are in the "IntRoutine" routine, but the free running counter is still running toward its next overflow meaning that whatever we do has to get done in less than 1/61 seconds if we are not going to miss the next interrupt, unless we make some other arrangements to count all the interrupts (with an internal subroutine or some other scheme). It can become quite complicated if a lot needs to be done, so we will not worry about it here.

Before going any further, let's take a closer look at the OPTION_REG and the INTCON (interrupt control) register. These are 8-bit registers with the 8 bits of each register assigned as follows:

OPTION_REG the option register:

Bit 7   RBPU. Not of interest to us at this time. (This bit enables the port B pull ups.)

Bit 6   INTEDG. Not of interest to us at this time. (The interrupt edge select bit determines which edge the interrupt will be selected on, rising [1] or falling [0]). Either one works for us.

Bit 5   T0CS, Timer0 clock select bit. Selects which clock will be used.
    1 = Transition on TOCKI pin.
    0 = Internal instruction cycle clock (CLKOUT). We will use this, the oscillator. See bit 4 description

Bit 4   T0SE, source edge select bit. Determines when the counter will increment.
    1 = Increment on high-to-low transition of TOCKI pin.
    0 = Increment on low-to-high transition of TOCKI pin.

Bit 3   PSA, prescalar assignment pin. Decides what the prescalar applies to.
    1 = Select watchdog timer (WDT)
    0 = Select Timer0. *We will be using this.*

Bits 2, 1, and 0 define the prescalar value for the timer. As mentioned earlier, the prescalar can be associated with Timer0 or with the watchdog timer (WDT) but not both. Note that the scaling for the WDT is half the value for Timer0 for the same three bits.

Bit value	TMR0 rate	WDT rate	
000	1:2	1:1	
001	1:4	1:2	
010	1:8	1:4	
011	1:16	1:8	
100	1:32	1:16	
101	1:64	1:32	*We will use this.*
110	1:128	1:64	
111	1:256	1:128	

Caution *A very specific sequence must be followed (which does not apply here) when changing the prescalar assignment from Timer0 to the WDT to make sure an unintended reset does not take place. This is described in detail in the PICmicro Mid-Range MCU Family Reference Manual (DS33023)*

As per the preceding, in our specific example, OPTION_REG is set to %10000101. Refer to the datasheet for more specific information.

INTCON the interrupt control register values are as follows:

Bit 7=1   Enables global interrupts.

Bit 6=1   Enables all peripheral interrupts.

Bit 5 =1   Enables an interrupt to be set on Bit 2 below when Timer0 overflows.

Bit 4 =1   Enables an interrupt if RB0 changes.

90 TIMERS AND COUNTERS

Bit 3 =1   Enables an interrupt if any of the PORTB pins are programmed as inputs and change state.

Bit 2   Is the Interrupt flag for Timer0.

Bit 1   Is the Interrupt flag for all internal interrupts.

Bit 0   Is the Interrupt flag for pins B7 to B4 if they change state.

Note that Bit 2 is set clear when we start and will be set to 1 when the first interrupt takes place. It has to be re-cleared within the interrupt service routine thereafter. (This is usually at the end of the routine, but not necessarily so.)

A TIMER0 CLOCK: FROM A PROGRAM BY MICROENGINEERING LABS (ON THEIR WEB SITE)

The following program, written by **microEngineering Labs** and provided by them as a part of the information on their Web site, demonstrates the use of interrupts to create a reasonably accurate clock that uses the LCD display to show the time in hours, minutes, and seconds.

LCD CLOCK PROGRAM USING ON INTERRUPT

This program uses TMR0 and prescalar. Watchdog timer should be set to OFF at program time, and Nap and Sleep should not be used.

Buttons may be used to set hours and minutes.

In Program 6.3, the CLEAR and OSC commands are not used, but we will always use them in our programs.

Program 6.3   Timer0 usage per microEngineering Labs program (Hours, seconds, and minutes digital clock)

```
DEFINE LCD_DREG PORTD      ; define LCD connections
DEFINE LCD_DBIT 4          ;
DEFINE LCD_RSREG PORTE     ;
DEFINE LCD_RSBIT 0         ;
DEFINE LCD_EREG PORTE      ;
DEFINE LCD_EBIT 1          ;
                           ;
HOUR VAR BYTE              ; define hour variable
DHOUR VAR BYTE             ; define display hour variable
MINUTE VAR BYTE            ; define minute variable
SECOND VAR BYTE            ; define second variable
TICKS VAR BYTE             ; define pieces of seconds variable
UPDATE VAR BYTE            ; define variable to indicate update of LCD
I VAR BYTE                 ; de bounce loop variable
ADCON1 = %00000111         ; parts of PORTA and E made digital
LOW PORTE.2                ; LCD R/W low = write
PAUSE 100                  ; wait for LCD to startup
```

(*Continued*)

Program 6.3 Timer0 usage per microEngineering Labs program (Hours, seconds, and minutes digital clock) (*Continued*)

```
HOUR = 0                         ; set initial time to 00:00:00
MINUTE = 0                       ;
SECOND = 0                       ;
TICKS = 0                        ;
UPDATE = 1                       ; force first display
                                 ; set TMR0 to interrupt every 16.384
                                 ; milliseconds
OPTION_REG = %01010101           ; set TMR0 configuration and enable
                                 ; PORTB pullups
INTCON = %10100000               ; enable TMR0 interrupts
ON INTERRUPT GOTO TICKINT        ;
                                 ; main program loop -
MAINLOOP:                        ; in this case, it only updates the LCD
                                 ; with the it
TRISB = %11110000                ; enable all buttons
PORTB =%00000000                 ; PORTB lines low to read buttons
                                 ; check any button pressed to set time
IF PORTB.7 = 0 THEN DECMIN       ;
IF PORTB.6 = 0 THEN INCMIN       ; last 2 buttons set minute
IF PORTB.5 = 0 THEN DECHR        ;
IF PORTB.4 = 0 THEN INCHR        ;
                                 ; first 2 buttons set hour
CHKUP:  IF UPDATE = 1 THEN       ; check for time to update screen
LCDOUT $FE, 1                    ; clear screen
                                 ; display time as hh:mm:ss
DHOUR = HOUR                     ; change hour 0 to 12
IF (HOUR // 12) = 0 THEN         ;
    DHOUR = DHOUR + 12           ;
ENDIF                            ;
                                 ;
IF HOUR < 12 THEN                ; check for AM or PM
  LCDOUT DEC2 DHOUR, ":", DEC2 MINUTE, ":", DEC2 second, " AM"_
  ELSE                           ;
    LCDOUT DEC2 (DHOUR - 12), ":", DEC2 MINUTE, ":", DEC2 SECOND,_
    " PM"
  ENDIF                          ;
  UPDATE = 0                     ; screen updated
ENDIF                            ;
GOTO MAINLOOP                    ; do it all forever
                                 ; increment minutes
INCMIN: MINUTE = MINUTE + 1 ;
IF MINUTE >= 60 THEN             ;
  MINUTE = 0                     ;
ENDIF                            ;
GOTO DEBOUNCE                    ;
                                 ; increment hours
```

(*Continued*)

Program 6.3 Timer0 usage per microEngineering Labs program (Hours, seconds, and minutes digital clock) (*Continued*)

```
INCHR:  HOUR = HOUR + 1        ;
IF HOUR >= 24 THEN             ;
  HOUR = 0                     ;
ENDIF                          ;
GOTO DEBOUNCE                  ;
                               ; decrement minutes
DECMIN: MINUTE = MINUTE - 1    ;
IF MINUTE >= 60 THEN           ;
  MINUTE = 59                  ;
ENDIF                          ;
GOTO DEBOUNCE                  ;
                               ; decrement hours
DECHR:  HOUR = HOUR - 1        ;
IF HOUR >= 24 THEN             ;
  HOUR = 23                    ;
ENDIF                          ;
                               ; de-bounce and delay for 250 ms
DEBOUNCE: FOR I = 1 TO 25      ;
  PAUSE 10                     ; 10 ms at a time so no interrupts
                               ; are lost
NEXT I                         ;
UPDATE = 1                     ; set to update screen
GOTO CHKUP                     ;
                               ; interrupt routine to handle each
                               ; timer tick
DISABLE                        ; disable interrupts during
                               ; interrupt handler
  TICKINT: TICKS = TICKS + 1   ; count pieces of seconds
    IF TICKS < 61 THEN TIEXIT  ; 61 ticks per second (16.384 ms
                               ; per tick)
                               ; one second elapsed - update time
    TICKS = 0                  ;
    SECOND = SECOND + 1        ;
      IF SECOND >= 60 THEN     ;
        SECOND = 0             ;
        MINUTE = MINUTE + 1    ;
      IF MINUTE >= 60 THEN     ;
        MINUTE = 0             ;
        HOUR = HOUR + 1        ;
          IF HOUR >= 24 THEN   ;
            HOUR = 0           ;
          ENDIF                ;
        ENDIF                  ;
      ENDIF                    ;
      UPDATE = 1               ; set to update LCD
TIEXIT: INTCON.2 = 0           ; reset timer interrupt flag
RESUME                         ;
END                            ;
```

In the preceding clock, the keyboard buttons are used as follows

SW1 and SW5 increment the hours.

SW2 and SW6 decrement the hours.

SW3 and SW7 increment the minutes.

SW4 and SW8 decrement the minutes.

The seconds cannot be affected other than with the reset switch.

Timer1: The Second Timer

The second timer, Timer1 is the 16-bit timer/counter. This is the most powerful timer in the MCU. As such, it is the hardest of the timers to understand and use but is also the most flexible of the three timers. It consists of two 8-bit registers and each register can be read and written to. The timer can be used either as a timer or as a counter depending on how the Timer1 clock select bit (TMR1CS), which is bit 1 in the Timer1 control register (T1CON), is set.

In Timer1, we can control the value that the timer starts its count with, and thus change the frequency of the interrupts. Here we are looking to see the effect of changing the value preload into Timer1 on the frequency of the interrupts as reflected in a very rudimentary pseudo-stopwatch. The higher the value of the preload, the sooner the counter will get to $FF and the faster the interrupts will come. We will display the value of the prescalar loaded into the timer on the LCD so we can see the correlation between the values and the actual operation of the interrupts. As the interrupts get closer and closer together, the time left to do the main task gets smaller and smaller and you can see this in the speed at which the stopwatch runs.

In the following program:

SW1 turns the stopwatch on.

SW2 stops the stopwatch.

SW3 resets the stopwatch.

POT0, the first potentiometer, is read and then written into TMR1H. (TMR1L is ignored in our case, but you may want to use it in a more critical application.)

The results of the experiment are shown in the LCD display.

Let's creep up on the solution. We will develop the program segments and discuss them as we go, putting the segments together later for a program we can run, as shown in Program 6.4.

Program 6.4 **Timer1 usage** (Rudimentary timer operation that depends on value of POT-1)

```
; first let us set up the LCD display parameters.
CLEAR                       ; clear memory
DEFINE OSC 4                ; set osc speed
DEFINE LCD_DREG PORTD       ; lcd is on PORTD
DEFINE LCD_DBIT 4           ; we will use 4-bit protocol
DEFINE LCD_RSREG PORTE      ; register select register
DEFINE LCD_RSBIT 0          ; register select bit
DEFINE LCD_EREG PORTE       ; enable Register
DEFINE LCD_EBIT 1           ; enable bit
PORTE.2 = 0                 ; set for write mode
PAUSE 500                   ; wait .5 seconds
                            ;
; Next let us define the variables we will be using
ADVAL VAR BYTE              ; create adval to store result
TICKS VAR WORD              ;
TENTHS VAR BYTE             ;
SECS VAR WORD               ;
MINS VAR BYTE               ;
                            ;
; Set the variable to specific values, not necessary in this program
; but a formality for clarity
TICKS = 0                   ;
TENTHS = 0                  ;
SECS = 0                    ;
MINS = 0                    ;
;
; Set the registers that will control the work. ;
; This is the nitty gritty of it so we will call out each bit.
; INTCON is the interrupt control register.
;
INTCON      =%11000000
; bit 7: GIE: Global Interrupt Enable bit, this has to be set
;   for any interrupt to work.
;           1 = Enables all un-masked interrupts
;           0 = Disables all interrupts
; bit 6: PEIE: Peripheral Interrupt Enable bit
;           1 = Enables all un-masked peripheral interrupts
;           0 = Disables all peripheral interrupts
; bit 5: T0IE: TMR0 Overflow Interrupt Enable bit
;           1 = Enables the TMR1 interrupt
;           0 = Disables the TMR1 interrupt
; bit 4: INTE: RB0/INT External Interrupt Enable bit
;           1 = Enables the RB0/INT external interrupt
;           0 = Disables the RB0/INT external interrupt
```

(*Continued*)

Program 6.4 Timer1 usage (Rudimentary timer operation that depends on value of POT-1) (Continued)

```
; bit 3:  RBIE: RB Port Change Interrupt Enable bit
;           1 = Enables the RB port change interrupt
;           0 = Disables the RB port change interrupt
; bit 2:  T0IF: TMR0 Overflow Interrupt Flag bit
;           1 = TMR0 register has overflowed (must be cleared in
;               software)
;           0 = TMR0 register did not overflow
; bit 1:  INTF: RB0/INT External Interrupt Flag bit
;           1 = The RB0/INT external interrupt occurred (must be
;               cleared in software)
;           0 = The RB0/INT external interrupt did not occur
; bit 0:  RBIF: RB Port Change Interrupt Flag bit
;           1 = At least one of the RB7:RB4 pins changed state (must
;               be cleared in software)
;           0 = None of the RB7:RB4 pins have changed state
;
; T1CON is the timer 1 control register.
T1CON       =%00000001
; bit 7-6: Unimplemented: Read as '0'
; bit 5-4: T1CKPS1:T1CKPS0: Timer1 Input Clock Prescale Select bits
;           11 = 1:8 Prescale value
;           10 = 1:4 Prescale value
;           01 = 1:2 Prescale value
;           00 = 1:1 Prescale value
; bit 3:  T1OSCEN: Timer1 Oscillator Enable Control bit
;           1 = Oscillator is enabled
;           0 = Oscillator is shut off (The oscillator inverter
;               is turned off to eliminate power drain)
; bit 2:  T1SYNC: Timer1 External Clock Input Synchronization
;           Control bit
;           TMR1CS = 1
;           1 = Do not synchronize external clock input
;           0 = Synchronize external clock input
;           TMR1CS = 0
;           This bit is ignored. Timer1 uses the internal clock
;           when TMR1CS = 0.
; bit 1:  TMR1CS: Timer1 Clock Source Select bit
;           1 = External clock from pin RC0/T1OSO/T1CKI (on the
;               rising edge)
;           0 = Internal clock (FOSC/4)
; bit 0:  TMR1ON: Timer1 On bit
;           1 = Enables Timer1
;           0 = Stops Timer1
;
; The option register
```

(Continued)

Program 6.4 Timer1 usage (Rudimentary timer operation that depends on value of POT-1) (*Continued*)

```
OPTION_REG = %00000000   ; Set Bit 7 to 0 and enable PORTB pullups
                         ; All other bits are for Timer 0 and
                         ; not applicable
                         ; here
PIE1=%00000001           ; See datasheet, enables interrupt.
ADCON0= %11000001        ; Configure and turn on A/D Module
; bit 7-6: ADCS1: ADCS0: ; A/D Conversion Clock Select bits
;           00 = FOSC/2
;           01 = FOSC/8
;           10 = FOSC/32
;           11 = FRC (clock derived from an RC oscillation)
; bit 5-3: CHS2:CHS0: Analog Channel Select bits
;           000 = channel 0, (RA0/AN0)
;           001 = channel 1, (RA1/AN1)
;           010 = channel 2, (RA2/AN2)
;           011 = channel 3, (RA3/AN3)
;           100 = channel 4, (RA5/AN4)
;           101 = channel 5, (RE0/AN5)(1)
;           110 = channel 6, (RE1/AN6)(1)
;           111 = channel 7, (RE2/AN7)(1)
; bit 2: GO/DONE: A/D Conversion Status bit
;           If ADON = 1   See bit 0
;           1 = A/D conversion in progress (setting this bit starts
;               the A/D conversion)
;           0 = A/D conversion not in progress (This bit is
;               automatically cleared by hardware when the A/D
;               conversion is complete)
; bit 1: Unimplemented: Read as ; 0;
; bit 0: ADON: A/D On bit
;           1 = A/D converter module is operating
;           0 = A/D converter module is shutoff and consumes no
;               operating current
;
; The A to D control Register for Port A is ADCON1
ADCON1 = %00000010   ; set part of PORTA analog
; The relevant table is on page 112 of the datasheet
; There are a number of choices which give us analog capabilities
; on PORTA.0
; and allow the voltage ; reference between Vdd and Vss. We have
; chosen 0010
; on the third line down in the table
;
; Next let us set up the port pin directions
TRISA = %11111111    ; set PORTA to all input
TRISB = %11110000    ; set up PORTB for keyboard reads
PORTB.0 = 0          ; set so we can read row 1 only for now
                     ;
```

(*Continued*)

Program 6.4 Timer1 usage (Rudimentary timer operation that depends on value of POT-1) (Continued)

```
ON INTERRUPT GOTO TICKINT   ; tells the program where to go on
                            ; interrupt
                            ;
                            ; Initialize display and write to top
                            ; line
LCDOUT $FE, 1, $FE, $80, "MM SS T"  ;
                            ;
MAINLOOP:           ;
ADCON0.2 = 1  ; conversion to reads POT-1. Conversion start
;
              ; now and takes place during loop. If loop was
              ; short we would allow for that.
              ; then check the buttons to decide what to do
IF PORTB.4 = 0 THEN STARTCLOCK  ;
IF PORTB.5 = 0 THEN STOPCLOCK   ;
IF PORTB.6 = 0 THEN CLEARCLOCK  ;
                            ;
                            ; and display what the clock
                            ; status is
LCDOUT $FE, $80, DEC2 MINS, ":",DEC2 SECS, ":", DEC TENTHS,_
"POT1=", DEC ADVAL, " "

; We are now ready to read what potentiometer setting is.
ADVAL = ADRESH              ; we assumed that enough time has
                            ; passed
                            ; to have an updated value in the
                            ; registers. If not add wait
GOTO MAINLOOP               ; do it again
                            ;
DISABLE                     ; disable interrupts during
                            ; interrupt handler
TICKINT:                    ;
  TICKS = TICKS + 1         ; ticks are influenced by the setting
                            ; of POT-1
  IF TICKS < 5 THEN TIEXIT  ; arbitrary value to get one
                            ; second
  TICKS = 0                 ;
                            ;
  TENTHS = TENTHS + 1       ;
  IF TENTHS <9 THEN TIEXIT  ;
  TENTHS = 0                ;
                            ;
  SECS = SECS + 1           ; update seconds
  IF SECS < 59 THEN TIEXIT  ;
  SECS = 0                  ;
  MINS = MINS + 1           ; update minutes
  TIEXIT:                   ;
```

(*Continued*)

Program 6.4 **Timer1 usage** (Rudimentary timer operation that depends on value of POT-1) (*Continued*)

```
   IF PORTB.5 = 0 THEN STOPCLOCK ;
   TMR1H=ADRESH              ;
   PIR1=0                    ;
RESUME                       ; go back to the main routine
ENABLE                       ;
                             ;
DISABLE                      ;
STARTCLOCK:                  ;
   INTCON = %10100011        ; enable TMR1 interrupts
   TICKS = 0                 ;
GOTO MAINLOOP                ;
                             ;
STOPCLOCK:                   ;
   INTCON = %10000011        ; disable TMR1 interrupts
   PAUSE 2                   ;
   TICKS = 0                 ;
GOTO MAINLOOP                ;
                             ;
CLEARCLOCK:                  ;
   INTCON = %10000011        ; disable TMR1 interrupts
   MINS = 0                  ;
   SECS = 0                  ;
   TENTHS = 0                ;
   TICKS = 0                 ;
GOTO MAINLOOP                ;
ENABLE                       ;
END                          ;
```

Run this program to see how the setting of the potentiometer affects the operation of the stopwatch. It becomes clear that choosing how the interrupt will serve our purposes is very important, and a bad choice can pretty much compromise the operation of the program.

We can read the timer and the interrupts at our discretion either before or after an interrupt has occurred, and the interrupt flag can be cleared whenever we wish, if it has been set. If it has not been set, there is no need to clear it.

Figure 6.1 provides a diagrammatic representation of how an interrupt routine is implemented in a typical program.

Even the 16-bit Timer1 on the 16F877A cannot time a long interval. Repeated intervals have to be put together to create long time periods. The longest possible time between interrupts for Timer1 (with a 4-MHz clock) is 0.524288 seconds. The maximum prescale value is 1:8. The postscalar is only available on Timer2 (which in any case is a shorter 8-bit timer). This results in a maximum time and is determined by multiplying the instruction clock cycle (1 μsec @ 4 MHz) by the prescale (8) by the number of counts from one overflow to the next (65536). 1 μsec * 8 * 65536 = 0.524288 seconds. On a 20-MHz machine, the interval would be one-fifth of this.

TIMER1: THE SECOND TIMER

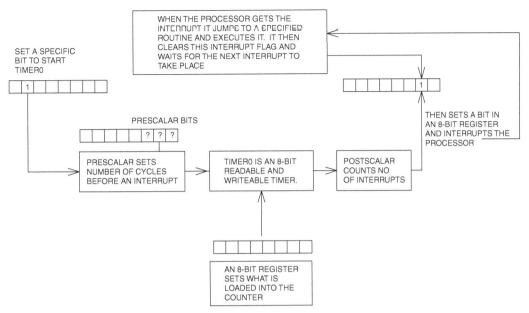

Figure 6.1 The simplified, basic structure of a typical interrupt routine. (Bits shown as being set are not the real bits.)

Timer1 uses two registers TMR1H and TMR1L.
The timer has the following general properties:

1. Increments from $0000 to $FFFF in two registers.
2. If the interrupt is enabled, an interrupt will occur when the two byte counter overflows from $FFFF to $0000.
3. The device can be used as a timer.
4. The device can be used as a counter.
5. The timer registers can be read and written to at any time.
6. There is no postscalar for this timer.

Simply stated again, this timer is used by setting its registers to a specific value and using the interrupts this value creates in a useful way. A 16-bit timer will count up from where set to FFFF and then flip to the selected value and start over again. An interrupt occurs and the interrupt flag is set every time the register overflows from FFFF to 0. We respond to the interrupt by doing whatever needs to be done in response to the interrupt, resetting the interrupt flag and then going back to the main routine. On timers that permit the use of a prescalar and postscalars, the pre/postscalar allows us to increase the time between interrupts by multiplying the time between interrupts with a defined value in a 3- to 8-bit location. On writable timers, we have the ability to start the timers with values of our choice in the timer register(s). This gives us very useable but not absolute control over the interrupt intervals.

Consider the fact that a 0.01 second timer setting with a prescalar set to 16 would provide us with an interrupt every 0.16 seconds and we would have 0.16 seconds to do whatever we wanted to do between interrupts. Actually, less than 0.16 seconds, because there are still the other lines of code in that program that need to be executed.

So there are serious limits as to what can be put in the timer counter and what can be put in the prescalar. In addition, the interrupt frequency is (also) affected by the accuracy of the processor clock oscillator.

PRESCALARS

The value of the scaling that will be applied to the timer is determined by the contents of two bits in the interrupt control register. These bits multiply the time between interrupts by powers of 2 as explained in the following.

Prescalar	For Timer1		For the Watchdog Timer	
00	Multiply by	1	00	Multiply by	2
01	Multiply by	2	01	Multiply by	4
10	Multiply by	4	10	Multiply by	8
11	Multiply by	8	11	Multiply by	16

The two bits are bit 4 and bit 5 of the Timer1 control register T1CON.
The 8 bits in T1CON are assigned as follows

TMR1ON	Bit 0	1=Enables Timer1	0=Disables timer
TMR1CS	Bit 1	1=Use external clock	0=Use internal clock
TISYNC	Bit 2	1=Sync with internal clock input	0=Sync with external clock input
TIOSCEN	Bit 3	1=Enable oscillator	0=Shut off oscillator
T1CKPS1	Bit 4	*Counter scalar is described above*	
T1CKPS0	Bit 5	*Counter scalar is described above*	
~~———~~	~~Bit 6~~	~~Not used~~	
~~———~~	~~Bit 7~~	~~Not used~~	

USING TIMER1 TO RUN A CRITICAL INTERRUPT-DRIVEN TASK WHILE THE MAIN PROGRAM RUNS A FOREGROUND TASK

Let's use this timer in the same way we used Timer0 earlier and see what the differences between the two timers are. To begin with, because Timer1 is 16 bits wide, it can take much longer for it to set its interrupt flag. The interrupt flag was set approximately 61 times a second by Timer0. Timer1 flag can take approximately 0.524 seconds, as calculated earlier, so it is set about two times a second. Let's write a short Timer1 program that is similar to the original Timer0 blinker program to see how the differences shape up.

Program 6.5 blinks the LEDs at D0 and D1 ON and OFF alternately as the foreground part of the program. The interrupts generated by Timer1 are used to blink D3 ON and OFF at half-second intervals. Since the control of D3 is driven by the interrupt, the timing stays accurate. Any time used by the interrupt routine is lost by the foreground task and affects the overall frequency of the D0/D1 blink rate. It is important to understand this loss.

Program 6.5 Using Timer0 (Programs blinks two LEDs alternately and blinks a third LED approximately a half second ON and a half second OFF)

```
CLEAR                      ; clear
DEFINE OSC 4               ; osc speed
TRISD = %00000000          ; set all PORTD lines to output
TRISE = %00000000          ; set all PORTE lines to output
                           ; set the A to D control register for
                           ; digital ports D, E
ADCON1=%00000111           ; needed, 16F877A because it
                           ; has analog properties
T1CON = %00000001          ; turn on Timer0, prescalar = 1
INTCON = %11000000         ; enable global interrupts, peripheral
                           ; interrupts
                           ;
I VAR WORD                 ; counter variable
J VAR WORD                 ; counter variable
PAUSE 500                  ;
I=0                        ; set counters to 0
J=0                        ;
PIE1.0 = 1                 ; enable TMR1 overflow interrupt
ON INTERRUPT GOTO INTHANDLER  ;
PORTD=0                    ; turn off the entire port
PORTD.3 = 0                ; light d3 on bargraph off, repeats above instr
PORTD.2 = 0                ; light d3 on bargraph off, repeats above instr
                           ;
MAINLOOP:                  ;
IF PORTD.1 = 0 THEN        ; routine lights D0 and D1 alternately to
   PORTD.1 = 1             ; that the program is running the main routine
   PORTD.0 = 0             ;
ELSE                       ;
   PORTD.1 = 0             ;
   PORTD.0 = 1             ;
ENDIF                      ;
FOR I = 1 TO 300           ; this is in lieu of a long pause instruction
   PAUSEUS 100             ; so that interrupt is not compromised
NEXT I                     ;
GOTO MAINLOOP              ; do it all forever
                           ;
DISABLE                    ;
```

(Continued)

Program 6.5 Using Timer0 (Programs blinks two LEDs alternately and blinks a third LED approximately a half second ON and a half second OFF) (*Continued*)

```
INTHANDLER:                    ; this is the interrupt service routine
IF J < 6 THEN                  ; this routine allows 6
  J = J+1                      ; interrupts for each change of state
  GOTO COUNTNOTFULL            ; of LED D3.
ELSE                           ;
  J = 0                        ;
ENDIF                          ;
IF PORTD.3 = 1 THEN            ; the D3 blink routine
  PORTD.3 = 0                  ;
ELSE                           ;
  PORTD.3 = 1                  ;
ENDIF                          ;
COUNTNOTFULL:                  ;
PIR1.0 = 0                     ; must now clear the interrupt flag
RESUME                         ;
ENABLE                         ;
END                            ; end program
```

Play with the value of the counter *J* to see how this affects the operation of the program. Study the differences between the programs to set and clear the timer flags. Though both of the preceding programs do the same thing, the setting of the potentiometer in the first program must be modified to match the needs of the timer being used.

Timer2: The Third Timer

Timer2 is an 8-bit "timer only," meaning it cannot be used as a counter. It has a prescalar and a postscalar. The timer register for this counter is both writable and readable. If you can write to a counting register, you can set the value the count starts at and thus control the interval between interrupts (to some degree). That, and the ability to set the pre- and postscalars, gives you the control you need for effective control of the interrupt interval even though you still cannot time all events exactly because of the coarseness of the settings available. Timer2 has a period register PR2, which can be set by the user. The timer counts up from $00 to the value set in PR2, and when the two are the same, it resets to 0. Small values in PR2 can be used to create very rapid interrupts, so much so that there may be no time left to do anything else.

The Timer2 control register is T2CON and its 8 bits are assigned as follows:

T2CKPS0	Bit 0	Counter prescalar	0 = Disables timer
T2CKPS1	Bit 1	Counter prescalar	0 = Use internal clock
TMR2ON	Bit 2	1 = Timer2 is on	0 = Timer2 is off, shuts off oscillator

TOUTPS0	Bit 3	) Counter postscalar value
TOUTPS1	Bit 4	) Counter postscalar value
TOUTPS2	Bit 5	) Counter postscalar value
TOUTPS3	Bit 6	) Counter postscalar value
~~	~~Bit 7~~	~~Not used~~

As always, the input clock for this timer is divided by 4 before it is fed to the timer. On a processor running at 4 MHz, the feed to the timer is at 1 MHz.

Prescalar for Timer2

00	Multiply by	1, no scaling
01	Multiply by	4
1×	Multiply by	16

Postscalar for Timer2

0000	Multiply by	1, no scaling
0001	Multiply by	2
0010	Multiply by	3
0011	Multiply by	4
0100	Multiply by	5
0101	Multiply by	6
0110	Multiply by	7
0111	Multiply by	8
1000	Multiply by	9
1001	Multiply by	10
1010	Multiply by	11
1011	Multiply by	12
1100	Multiply by	13
1101	Multiply by	14
1110	Multiply by	15
1111	Multiply by	16

The timer is turned ON by setting bit 2 in register T2CON (the Timer2 control register).

The interrupt for Timer2 is enabled by setting Bit 1 of PIE1 and the interrupt lets the program know that it has occurred by setting Bit 1 in PIR1. (Bit 0 in both these registers are for Timer1.)

Bit 7 (the global interrupt enable bit) of INTCON, the interrupt control register, enables all interrupts, including those created by Timer2. Bit 6 of INTCON enables all unmasked peripheral interrupts, and using this feature is one of the ways of awakening a sleeping MCU.

Timer2 can also control the operation of the two PWM signals that can be programmed to appear on lines C1 and C2 with the HPWM command in PICBASIC PRO. Since this one timer controls both lines simultaneously, they both have to have the same PWM *frequency*. However, the relative *width of the pulse* within each of the PWM signals during each cycle *does not have to be the same*.

Timer2 is also used as a baud rate clock timer for communications. See page 54 of the datasheet.

MAKING SURE A TIMER IS WORKING

If you are uncomfortable about getting a timer working, or knowing for sure that it is working, write a very short program in which the main loop displays a variable that is incremented in the interrupt routine. If you see the value of the variable going up, you know the program is going to, and returning from, the interrupt routine.

WATCHDOG TIMER

A watchdog timer is a timer that sets an interrupt that tells us that for some reason the program has hung up or otherwise gone awry. As such, it is expected that in a properly written program, the watchdog timer will never set an interrupt. This is accomplished by resetting the watchdog timer every so often in the program. The compiler inserts these instructions automatically if the watchdog timer option is selected. However, setting the option does not guarantee that a program cannot or will not hang up. Software errors and infinite loops that reset the timer within themselves can still cause hangups.

The watchdog timer is scalable. It shares its scalar with Timer0 on an exclusive basis. Either it uses the scalar or Timer0 uses it. They cannot both use it at the same time. See discussion under Timer0 in the datasheet for more information.

Since PICBASIC PRO assumes that the watchdog timer will be run with a 1:128 prescalar, unwanted watchdog resets could occur when you assign the prescalar to Timer0. If you change the prescalar setting in OPTION_REG, you should disable the watchdog timer when programming. The watchdog enable/disable can be found on the configuration screen of your (hardware) programmer's software.

Counters

Of the three timers in the 16F877A, only Timer0 (the 8-bit timer) and Timer1 (the 16-bit timer) can be used as counters. Timer2 does not have a counter input line provided for it. Generally speaking, this makes Timer0 suitable for use with small counts and rapid interrupts, and Timer1 suitable for larger counts.

HOW DOES A COUNTER WORK?

The operation of a counter is similar to the operation of a timer except that instead of getting its count from an internal clock or oscillator, the counter gets its signals from an outside source. This means we have to do the following things to use a counter:

- Decide which counter (timer) to use.
- Tell the counter where the signal is coming from.
- Tell it whether to count on a rising or falling edge.
- Decide what target count we are looking for.
- Tell the counter where to start counting because the interrupt will occur when the counter gets full.
- Decide whether we will need to scale the count by setting the scalar(s).

Once we start a counter, the counting continues until we deactivate it. There is no other way to stop it, nor any reason for doing so. It will reset if the MCU is reset and it can be reset by writing to it. The rest has to do with knowing what bits to set in the control registers to get the counters to operate in the way we want them to.

USING TIMER0 AS A COUNTER

Note *Though often called TIMER0, and referred to as TIMER0 here and in the datasheet, the real designation of this timer address is TMR0.*

The three registers related to the control of the TIMER0 module are TMR0, INTCON, and OPTION_REG. INTCON is the interrupt control register. See the datasheet for more information.

This counter has the following properties:

- 8-bit timer/counter
- Readable and writable
- 8-bit software programmable prescalar
- Internal or external clock select
- Interrupt on overflow from FFh to 00h
- Edge select for external clock

Counter mode is selected by setting OPTION_REG.5=1.

- The external input for the timer will come in on PORTA.4, which is pin 6 on the PIC.
- The edge direction is selected in OPTION_REG.4 (1=Rising edge).
- The prescalar is assigned with bits 0 to 3 of the OPTION_REG register. (See page 49 of the datasheet.)

Note again that the watchdog timer *cannot* use the prescalar when the prescalar is being used by Timer0.

Since this is an 8-bit counter, it is suited to the counting of small number of counts, but longer counts can be accommodated by using a routine to keep track of the interrupts.

Let's use the counter to count the pulses received from a 20-slot encoder mounted on a small DC motor. This same source will be used later for the Timer1 experiment for comparison between counters.

We will set up to use the LCD display so we can display certain registers during the operation of Program 6.6. We will also set up to read the potentiometers so we can use their values to modify the program as it runs. Only POT0 and POT1 are used in the program.

Program 6.6 Using Timer0: Program counts the pulses from a motor-driven encoder (You can change the speed of the motor and the time for counts with the 2 POTs)

```
CLEAR
DEFINE OSC 4                  ; 4 MHz clock
DEFINE LCD_DREG PORTD         ; data register
DEFINE LCD_RSREG PORTE        ; register select
DEFINE LCD_RSBIT 0            ; pin number
DEFINE LCD_EREG PORTE         ; enable register
DEFINE LCD_EBIT 1             ; enable bit
DEFINE LCD_RWREG PORTE        ; read/write register
DEFINE LCD_RWBIT 2            ; read/write bit
DEFINE LCD_BITS 8             ; width of data
DEFINE LCD_LINES 2            ; lines in display
DEFINE LCD_COMMANDUS 2000     ; delay in micro seconds
DEFINE LCD_DATAUS 50          ; delay in micro seconds
                              ;
DEFINE CCP1_REG PORTC         ; define the hpwm settings
DEFINE CCP1_BIT 2             ;
                              ; define the A2D values
DEFINE ADC_BITS 8             ; set number of bits in result
DEFINE ADC_CLOCK 3            ; set internal clock source (3=rc)
DEFINE ADC_SAMPLEUS 50        ; set sampling time in us
                              ; set the analog to digital control
                              ; register
ADCON1=%00000110              ; needed for the 16F877A LCD
TEST VAR WORD                 ;
ADVAL0 VAR BYTE               ; create adval to store result
ADVAL1 VAR BYTE               ; create adval to store result
X VAR WORD                    ;
Y VAR WORD                    ;
PAUSE 500                     ; LCD start up
LCDOUT $FE, 1                 ; clear display
OPTION_REG=%00110000          ;
TMR0=0                        ;
                              ; set up the register i/o
TRISC = %11110001             ; PORTC.0 is going to be the input to
                              ; start the motor in that we are using
                              ; a motor
                              ; encoder for input
PORTC.3=0                     ; enable the motor
PORTC.2=0                     ; set the rotation direction
                              ;
LOOP:                         ;
ADCIN 0, ADVAL0               ; read channel 0 to ADVAL0
ADCIN 1, ADVAL1               ; read channel 1 to ADVAL1
ADCIN 3, ADVAL2               ; read channel 3 to ADVAL2
                              ;
TMR0=0                        ;
```

(Continued)

Program 6.6 Using Timer0: Program counts the pulses from a motor-driven encoder (You can change the speed of the motor and the time for counts with the 2 POTs) (*Continued*)

```
PAUSE ADVAL1              ;
X=TMR0                    ;
IF ADVAL0>20 THEN         ;
  HPWM 2, ADVAL0, 32000   ;
  LCDOUT $FE, $80, DEC4 X,"  ",DEC ADVAL1," "
  LCDOUT $FE, $C0, "PWM = ",DEC ADVAL0,"         "
ELSE                      ;
  LCDOUT $FE, $C0, "PWM TOO LOW ",DEC ADVAL0," "
ENDIF                     ;
                          ;
GOTO LOOP                 ;
END                       ;
```

Play with the values of the two potentiometers to see what happens. Be careful about overflowing the counter past 255. Unexpected results can appear.

The Timer0 counter is affected by the OPTION REGISTER bits as follows:

OPTION_REG.6 = 1   ; Interrupt on rising edge

OPTION_REG.5 = 0   ; External clock

OPTION_REG.4 = 1   ; Increment on falling edge not used

OPTION_REG.3 = 0   ; Assign prescalar to Timer0

OPTION_REG.2 = 1   ;] These 3 bits set the prescalar

OPTION_REG.1 = 1   ;] You can experiment with changing these 3

OPTION_REG.0 = 1   ;] bits to see what happens.

Put these and other values in the program and run the program. See what happens.

USING TIMER1 AS A COUNTER

The operation of Timer1 as a counter is similar to the operation of Timer0, but because Timer1 is a 16-bit timer, much longer counts can be handled, and counts coming in at faster rates can be counted. It also means that a lot more can be done in the TimerLoop and BlinkerLoop routines if the program is designed to do so. However, the setup for Timer1 is more complicated because of the more numerous options available.

The differences between the use of the two timers have to do with the setup of the controlling registers. Timer1 is controlled by/uses six registers as compared to three for Timer0. They are:

INTCON   Interrupt control register

PIR1   Peripheral interrupt register 1

PIE1	Peripheral interrupt enable register 1
TMR1L	Low byte of the timer register
TMR1H	High byte of the timer register
T1CON	Timer1 interrupt control

Again, the frequency of the oscillator is divided by 4 before being fed to the counter when you use the internal clock (Fosc/4).

Page 52 of the datasheet reads:

Counter mode is selected by setting bit TMR1CS. In this mode, the timer increments on every rising edge of clock input on pin RC1/T1OSI/CCP2, when bit T1OSCEN is set, or on pin RC0/T1OSO/T1CKI, when bit T1OSCEN is cleared.

So three of the pins on the 16F877A can be used as inputs to the Timer1 counter module. They are:

Pin PORTA.4   which is the external clock input. Pin 6 on the PIC.

Pin PORTC.0   selected by setting T1OSCEN =1

Pin PORTC.1   selected by setting T1OSCEN=0

Timer1 is enabled by setting T1CON.0=1. It stops when this bit is turned off or disabled.

The clock that Timer1 will use is selected by T1CON.1. The external clock is selected by setting this to 1. The input for this external clock must be on PORTB.4.

In summary, 8 bits in the Timer1 control register, T1CON, provides the following functions:

~~Bit 7   Not used and is read as a 0~~

~~Bit 6   Not used and is read as a 0~~

Bit 5   Input prescalar

Bit 4   Input prescalar

Bit 3   Timer1 oscillator enable

Bit 2   Timer1 external clock synchronization

Bit 1   Timer1 clock select

Bit 0   Timer1 enable

If the interrupts are not going to be used, the other registers can be ignored. Set T1CON = %00110001.

The setting of these bits is described in detail on page 51 of the datasheet. Let's look at Program 6.7, which reflects the preceding information.

; First let us define all the defines that we will need.
; Here all the defines are included as an example but
; not all are needed when using the LAB-X1.

Program 6.7   **Timer1 as counter** (Timer1 counts signals from a motor encoder)

```
CLEAR                           ; clear memory
DEFINE OSC 4                    ; 4 MHz clock
DEFINE LCD_DREG PORTD           ; data register
DEFINE LCD_RSREG PORTE          ; register select
DEFINE LCD_RSBIT 0              ; pin number
DEFINE LCD_EREG PORTE           ; enable register
DEFINE LCD_EBIT 1               ; enable bit
DEFINE LCD_RWREG PORTE          ; read/write register
DEFINE LCD_RWBIT 2              ; read/write bit
DEFINE LCD_BITS 8               ; width of data
DEFINE LCD_LINES 2              ; lines in display
DEFINE LCD_COMMANDUS 2000       ; delay in micro seconds
DEFINE LCD_DATAUS 50            ; delay in micro seconds
```

; The next two lines define which pin is going to be used for the
; HPWM signal
; that will control the speed of the motor. The encoder that we
; are looking
; at is attached to the motor

```
DEFINE CCP1_REG PORTC           ; define the HPWM settings
DEFINE CCP1_BIT 2               ; pin C1
```

; The next few lines define the reading of the three
; potentiometers on the board. Only the first
; potentiometer is being used in the program but the others are
; defined so that you can
; use them when you modify the program. The potentiometers give
; you values you can
; change in real time.
; define the A2D values

```
DEFINE ADC_BITS     8           ; set number of bits in result
DEFINE ADC_CLOCK    3           ; set internal clock source (3=rc)
DEFINE ADC_SAMPLEUS 50          ; set sampling time in uS
```

; Next we set ADCON1 to bring the MCU back into digital mode.
; Since this PIC has analog capability, it comes up in
; analog mode after a reset or on startup.
                                ; set the Analog-to-Digital control register
```
ADCON1=%00000111                ; needed for the LCD operation
                                ; we create the variables that we will need.
TMR1 VAR WORD                   ; set the variable for the timer
ADVAL0 VAR BYTE                 ; create adval to store result
ADVAL1 VAR BYTE                 ; create adval to store result
ADVAL2 VAR BYTE                 ; create adval to store result
```

(*Continued*)

Program 6.7   **Timer1 as counter** (Timer1 counts signals from a motor encoder)
(*Continued*)

```
X VAR WORD              ; spare variable for experimentation
Y VAR WORD              ; spare variable for experimentation
PAUSE 500               ; pause for LCD to start up
LCDOUT $FE, 1           ; clear Display and cursor home
                        ;
                        ; set up the register I/O
TRISC = %11110001       ; PORTC.0 is going to be the Input
CCP1CON = %00000101     ; capture every rising edge
T1CON = %00000011       ; no prescale/Osc off/Sync on
                        ; external source/TMR1 on
                        ; start the motor, using a motor encoder
                        ; for input
PORTC.3=0               ; enable the motor
PORTC.2=1               ; set the rotation direction
; Next we go into the body of the program. The loop starts with
; reading all
; three potentiometers though we are using only the first one to
; set the power
; and thus the speed of the motor.
LOOP:                   ;
ADCIN 0, ADVAL0         ; read channel 0 to ADVAL0
ADCIN 1, ADVAL1         ; read channel 1 to ADVAL1
ADCIN 3, ADVAL2         ; read channel 3 to ADVAL2
; If the duty cycle of the motor is less than 20 out of 255 the
; motor will not come on
; so we make an allowance for that and display the condition on
; the LCD.
IF ADVAL0>20 THEN       ;
  HPWM 2, ADVAL0, 32000 ;
  LCDOUT $FE, $C0, "PWM = ",DEC ADVAL0,"         "
ELSE                    ;
  LCDOUT $FE, $C0, "PWM TOO LOW ",DEC ADVAL0,"   "
ENDIF                   ;
; Then we read the two timer registers to see how many counts
; went by.
; In our case the counts were too low to show up in the high
; bits so that
; it can be ignored but if you have a faster count input, you
; might want
; to add this information to the readout.
TMR1H = 0               ; clear Timer1 high 8-bits
TMR1L = 0               ; clear Timer1 low 8-bits
T1CON.0 = 1             ; start 16-bit timer
PAUSE 100               ; capture 100 ms of Input Clock Frequency
```

(*Continued*)

Program 6.7   **Timer1 as counter** (Timer1 counts signals from a motor encoder) *(Continued)*

```
T1CON.0 = 0              ; stop 16-bit Timer
TMR1.BYTE0 = TMR1L       ; read Low 8-bits
TMR1.BYTE1 = TMR1H       ; read High 8-bits
TMR1 = TMR1 - 11         ; capture Correction
IF TMR1 = 65525 THEN NOSIGNAL  ; see PICBASIC PRO manual for
                               ; explanation.
  LCDOUT $FE, $80, DEC5 TMR1," COUNTS"  ; frequency display
  PAUSE 10               ; slow down
GOTO LOOP                ; do it again
                         ;
NOSIGNAL:                ;
  LCDOUT $FE, $80, "NO SIGNAL " ;
GOTO LOOP                ;
END                      ;
```

PRESCALARS AND POSTSCALARS

Prescalars and postscalars can be confusing for the beginner. Here is a simple explanation.

A prescalar is applied to the system clock and affects the timer by slowing down the system clock as it applies to the timer. Normally, the timer is fed by a fourth of the basic clock frequency, which is called Fosc/4. In a system running a 4 MHz clock, the timer sees a clock running at 1 MHz. If the prescalar is set for 1:8, the clock will be slowed down by another eight times, and the timer will see a clock at 125 KHz. Refer to the diagram on the bottom half of page 52 (for Timer1) in the datasheet to see how this applies to Timer1.

A postscalar is applied after the timer count exceeds its maximum value, generating an overflow condition. The postscalar setting determines how many overflows will go by before an interrupt is triggered. If the postscalar is set for 1:16, the timer will overflow 16 times before an interrupt flag is set. The upper diagram on page 55 (for Timer2) of the datasheet shows this in its diagrammatic form and is worth studying.

All other things being equal, both pre- and postscalars are used to increase the time between interrupts.

When starting out, just leave the scalars at 1:1 values and nothing will be affected. We will not need to use them for any of the experiments we will be doing. Once you get more sophisticated in the use of timers, you can play with the values and learn more about how to use them. The primary use is in creating accurate timing intervals for communications and so on, because no external routines are necessary when this is done with scalars. When doing it this way, everything becomes internal to the PIC and is therefore not affected by external disturbances.

Additional information on timer modules is available in the *PICmicro Mid-Range MCU Family Reference Manual* (DS33023).

Exercises for Timers

1. Write a program to generate a 1-minute timer/clock with a 0.1 second display. Check its accuracy with the time site on the Internet.
   Make adjustments so it is accurate to within 1 second per hour, and then 1 second per day. Can this be done? Why? Which timer works best? Which timer is the easiest to use for such a task?

2. Write the preceding program for each of the other two timers.

Exercises for Counters

1. Design and make a tachometer for small model aircraft engines. Have a range of from 5 *rev per second* to 50,000 *rev per minute* displayed on the LCD in real-time.

2. Design and build a thermometer based on the changes in frequency exhibited by a 555 timer circuit being controlled by a thermistor. Calibrate the thermometer with a lookup table. If you are not familiar with the use of lookup tables, you should undertake the research necessary to understand how to use them. They are very useful devices at the level we are working.

7

CLOCKS, MEMORY, AND SOCKETS

Sockets U3, U4, and U5: For Serial One-Wire Memory Devices

Most PIC microcontrollers come with a certain amount of on-chip memory. This memory is enough for most applications created for these tiny processors, but there are times when more memory is needed to get the job done. The LAB-X1 has five empty 8-pin sockets. Three of these (the three on the left) are designed to allow us to experiment with three types of single-wire memory ICs. The ICs don't need just one wire for full control, but the data does go back and forth on one wire.

Note *Each memory socket accepts only one type of memory device, and only one of the ICs is allowed to be in place at any one time because the lines are shared between the sockets, so having more than one device plugged in can create conflicts.*

Depending on the type of memory you want to experiment with, one of the three schematics in Figure 7.1 is applicable.

The interfaces that have been developed for the three types of one-wire memory give you the choices you need for flexibility in board design and layout, but it also means that a single interface and protocol won't work for everything. The interfaces vary in speed, number of signal lines, and in other important details.

Since the memories are all one-wire serial devices, their memory content can vary from 128 bytes to 4 kilobytes or more and still maintain the 8-pin interface.

The salient characteristics of the three types of memory are as follows:

- **I2C SEEPROM** I2C SEEPROMs are serial, electrically erasable and programmable, read-only memories. They are best suited for applications needing a modest amount of inexpensive nonvolatile memory where a lot of I/O lines are not available for memory transfers. Requires four lines for control.

114　CLOCKS, MEMORY, AND SOCKETS

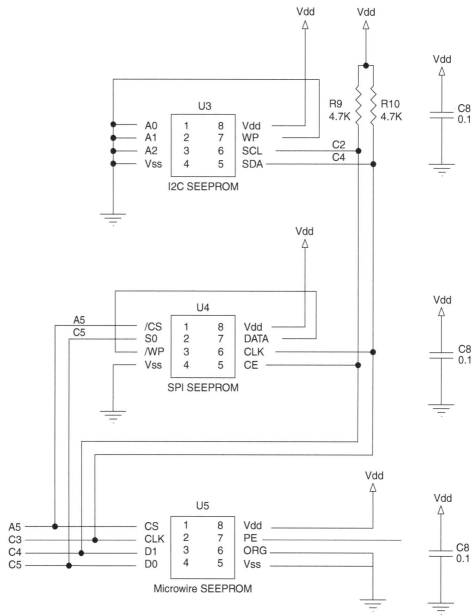

Figure 7.1 **One-wire memory sockets.** (The three types of memory you can experiment with on the LAB-X1 and their wiring layouts; only one IC may be in place at any one time.)

- **Microwire** Microwire is a National Semiconductor standard and is specially suited to use with their microcontrollers. Though often called a three-wire interface, it is actually a five-wire interface with four signal lines and a ground.
- **SPI** SPI (serial peripheral interface) originated at Motorola. It is much like Microwire, though the signal names, polarities, and other details vary. Like Microwire, SPI is often referred to as a three-wire interface, though a read/write interface actually requires two data lines, a clock, a chip select, and a common ground, making five wires.

Other manufacturers provide products to meet the standards that have been established.

Which EEPROM Type Should You Use?

I2C is the best EEPROM type to use if you have just two signal lines to spare, or if you have a cabled interface (I2C also has the strongest drivers).

If you want a clock rate faster than 400 kHz, use Microwire or SPI.

For more on using serial EEPROMs, refer to the manufacturers' pages on the Web, especially the following sites:

- National Semiconductor
    www.national.com/design/
    (Contains many application notes on Microwire)
- Motorola Semiconductor
    www.mcu.motsps.com/mc.html
    (Onsite microcontroller references contain SPI documentation.)

Jan Axelson's article in *Circuit Cellar* is a good source of detailed information on these devices. The article can be found at www.lvr.com/files/seeprom.pdf.

The PICBASIC PRO Compiler provides the instructions necessary to access these serial memories.

In all the programs I have listed in this chapter, I have used the programs and documentation verbatim from the microEngineering Labs Web site. I did this for two reasons. One, to not reinvent the wheel since the work has already been done by microEngineering Labs, and two, to expose the readers to how the programs are structured by a programmer other than myself. Each programmer has his or her own style of doing things and it is well worth being exposed to the work of other programmers.

Socket U3—I2C SEEPROM

Socket U3 accommodates I2C memory only. Figure 7.2 and Program 7.1 illustrate the use of this memory type.

116 CLOCKS, MEMORY, AND SOCKETS

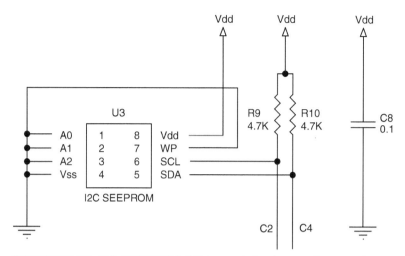

Figure 7.2 I2C SEEPROM. (Wiring and circuitry requirements.)

Program 7.1 Program to read from and write to I2C SEEPROMs

```
;This microEngineering Labs program can be found on their
;Web site. Use the latest version.
CLEAR                                          ; clear memory
SO CON 0                                       ; define serial output pin
N2400 CON 4                                    ; set serial mode
                                               ; define variables
DPIN    VAR PORTA.0                            ; I2C data pin
CPIN    VAR PORTA.1                            ; I2C clock pin
B0 VAR BYTE                                    ; variable declaration
B1 VAR BYTE                                    ; variable declaration
B2 VAR BYTE                                    ; variable declaration
                                               ; write to the memory
FOR B0 = 0 TO 15                               ; loop 16 times
   I2CWRITE DPIN, CPIN, $A0, B0, [B0]          ; write each location's
                                               ; address to itself
   PAUSE 10                                    ; delay 10 ms after each
                                               ; write is needed
NEXT B0                                        ;
                                               ;
LOOP:                                          ;
FOR B0 = 0 TO 15 STEP 2                        ; loop 8 times
   I2CREAD DPIN, CPIN, $A0, B0, [B1, B2]       ; read 2 locations in
                                               ; a row
   SEROUT SO, N2400, [#B1," ",#B2," "]         ; print 2 locations to
                                               ; CRT
NEXT B0                                        ;
                                               ;
SEROUT SO, N2400, [13,10]                      ; print a line feed
GOTO LOOP                                      ;
END                                            ;
```

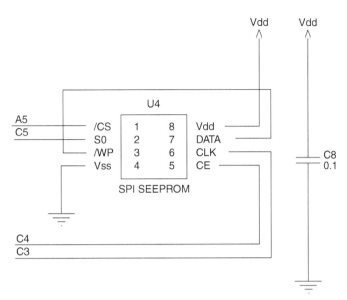

Figure 7.3 SPI SEEPROM. (Wiring and circuitry requirements.)

Socket U4—SPI SEEPROM

Socket U4 is wired to use SPI memory only. Figure 7.3 and Program 7.2 illustrate the use of this memory type.

Program 7.2 Program to read from and write to SPI SEEPROMs

```
; This microEngineering Labs program can be found on their
; Web site. Use the latest version.
; PICBASIC PRO program to read and write to SPI SEEPROMs
; Write to the first 16 locations of an external serial EEPROM
; Read first 16 locations back and send to LCD repeatedly
; Note: For SEEPROMs with word-sized addresses
DEFINE LOADER_USED 1            ; allows use of the boot
                                ; loader.
                                ; this will not affect
                                ; normal program operation.
DEFINE LCD_DREG PORTD           ; define LCD registers and
                                ; bits
DEFINE LCD_DBIT 4               ;
DEFINE LCD_RSREG PORTE          ;
DEFINE LCD_RSBIT 0              ;
DEFINE LCD_EREG PORTE           ;
DEFINE LCD_EBIT 1               ;
INCLUDE "MODEDEFS.BAS"          ;
```

(Continued)

Program 7.2 Program to read from and write to SPI SEEPROMs (*Continued*)

```
CS VAR PORTA.5                    ; chip select pin
SCK VAR PORTC.3                   ; clock pin
SI VAR PORTC.4                    ; data in pin
SO VAR PORTC.5                    ; data out pin
ADDR VAR WORD                     ; address
B0 VAR BYTE                       ; data
TRISA.5 = 0                       ; set CS to output
ADCON1 =%00000111                 ; set all of PORTA and PORTE
                                  ; to digital
LOW PORTE.2                       ; LCD R/W line low (W)
PAUSE 100                         ; wait for LCD to start up
FOR ADDR = 0 TO 15                ; loop 16 times
  B0 = ADDR + 100                 ; B0 is data for SEEPROM
  GOSUB EEWRITE                   ; write to SEEPROM
  PAUSE 10                        ; delay 10 ms after each
                                  ; write
NEXT ADDR                         ;
LOOP:                             ;
FOR ADDR = 0 TO 15                ; loop 16 times
  GOSUB EEREAD                    ; read from SEEPROM
  LCDOUT $FE, 1, #ADDR,": ",#B0   ; display
  PAUSE 1000                      ;
  NEXT ADDR                       ;
GOTO LOOP                         ;
; Subroutine to read data from addr in serial EEPROM
EEREAD: CS = 0                    ; enable serial EEPROM
  SHIFTOUT SI, SCK, MSBFIRST, [$03, ADDR.BYTE1, ADDR.BYTE0]
                                  ; Send read cmd and addr
  SHIFTIN SO, SCK, MSBPRE, [B0]   ; Read data
  CS = 1                          ; disable
RETURN                            ;
                                  ; Subr to write data at addr
                                  ; in serial EEPROM
EEWRITE:                          ;
  CS = 0                          ; enable serial EEPROM
  SHIFTOUT SI, SCK, MSBFIRST, [$06] ; send write enable command
  CS = 1                          ; disable to execute command
  CS = 0                          ; enable
  SHIFTOUT SI, SCK, MSBFIRST, [$02, ADDR.BYTE1, ADDR.BYTE0, B0]
                                  ; Sends address and data
  CS = 1                          ; disable
RETURN                            ;
END                               ; end program
```

Socket U5—Microwire Devices

Socket U5 is wired to use Microwire memory. Figure 7.4 and Program 7.3 illustrate the use of this memory type.

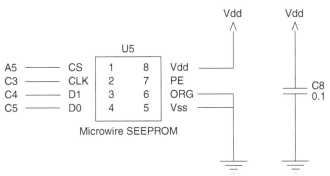

Figure 7.4 Microwire SEEPROM. (Wiring and circuitry requirements.)

This microEngineering Labs program (Program 7.3)—to read and write Microwire SEEPROM devices—can be found on their Web site. Use the latest version.

Program 7.3 Program to read from and write to Microwire SEEPROMs

```
; PICBASIC PRO program to read and write to Microwire
; SEEPROM 93LC56A
; Write to the first 16 locations of an external serial EEPROM
; Read first 16 locations back and send to LCD repeatedly
; Note: For SEEPROMs with byte-sized address
                                        ;
DEFINE LCD_DREG PORTD                   ; define LCD registers and bits
DEFINE LCD_DBIT 4                       ;
DEFINE LCD_RSREG PORTE                  ;
DEFINE LCD_RSBIT 0                      ;
DEFINE LCD_EREG PORTE                   ;
DEFINE LCD_EBIT 1                       ;
INCLUDE "MODEDEFS.BAS"                  ;
CS VAR PORTA.5                          ; chip select pin
CLK VAR PORTC.3                         ; clock pin
DI VAR PORTC.4                          ; data in pin
DO VAR PORTC.5                          ; data out pin
ADDR VAR BYTE                           ; address
B0 VAR BYTE                             ; data
LOW CS                                  ; chip select inactive
ADCON1 = 7                              ; set PORTA and PORTE to digital
LOW PORTE.2                             ; LCD R/W line low (W)
PAUSE 100                               ; wait for LCD to start up
GOSUB EEWRITEEN                         ; enable SEEPROM writes
FOR ADDR = 0 TO 15                      ; loop 16 times
  B0 = ADDR + 100                       ; B0 is data for SEEPROM
  GOSUB EEWRITE                         ; write to SEEPROM
  PAUSE 10                              ; Delay 10ms after each write
```

(Continued)

Program 7.3 Program to read from and write to Microwire SEEPROMs (*Continued*)

```
NEXT ADDR                               ;
LOOP:                                   ;
FOR ADDR = 0 TO 15                      ; loop 16 times
  GOSUB EEREAD                          ; read from SEEPROM
  LCDOUT $FE, 1, #ADDR,": ",#B0         ; display
  PAUSE 1000                            ;
  NEXT ADDR                             ;
GOTO LOOP                               ;
; Subroutine to read data from addr in serial EEPROM
EEREAD:                                 ;
CS = 1                                  ; enable serial EEPROM
SHIFTOUT DI, CLK, MSBFIRST, [%1100\4, ADDR]   ; send read cmd
                                              ; and address
SHIFTIN DO, CLK, MSBPOST, [B0] ; read data
CS = 0                                  ; disable
RETURN                                  ;
                                        ; subroutine to write data at
                                        ; addr in serial EEPROM
EEWRITE: CS = 1                         ; Enable serial EEPROM
SHIFTOUT DI, CLK, MSBFIRST, [%1010\4, ADDR, B0]
                                        ; sends write command, address
                                        ; and data
CS = 0                                  ; disable
RETURN                                  ;
; Subroutine to enable writes to serial EEPROM
EEWRITEEN:                              ; subroutine
CS = 1                                  ; enable serial EEPROM
SHIFTOUT DI, CLK, MSBFIRST, [%10011\5, 0\7]
                                        ; send write enable cmd and
                                        ; dummy clocks
CS = 0                                  ; disable
RETURN                                  ;
END                                     ; end program
```

Socket U6—Real-Time Clocks

Four options are available for using socket U6. This socket is designed to let us experiment with three real-time clocks and with a 12-bit analog-to-digital converter. It connects to the microcontroller as shown:

As shown in Figure 7.5, this is a four-wire interface between the MCU and the IC. The wiring for this chip is similar to the wiring for the Microwire SEEPROMS and the Microwire memory ICs. Essentially, this looks like a memory chip to the processor. When we write to this memory, we are writing to the clock, and when we read from this chip, we are reading an ever changing memory content that gives us information that we can interpret as "time." So the program to read and write to this clock looks like a program that interacts with the Microwire family of SEEPROMS. (See Program 7.3.)

SOCKET U6—REAL-TIME CLOCKS

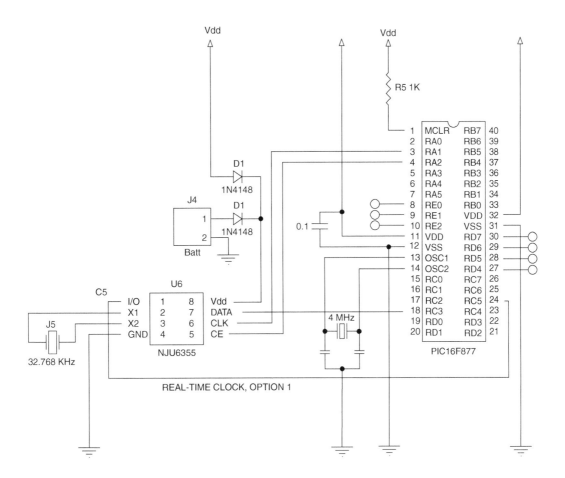

Figure 7.5 Clock implemented using IC NJU6355. (To the MCU, this clock IC looks like a set of memory locations.)

The same is true for the other chips. See Figures 7.6 and 7.7.

The NJU6355, the DS1202, and the DS1302 real-time clocks are the three integrated circuits for use in socket U6.

Note 1 *Jumper J5 which is used for soldering in the crystal for the clock ICs is also the connection that the analog signal for the 12-bit A-to-D converter goes into. So, if you solder in a crystal, you will have to remove the crystal and make arrangements to read in the analog signal when you want to experiment with the LTC1298 12-bit A-to-D converter. The A-to-D converter uses the same socket (U6) as is used by the three clock chips.*

122 CLOCKS, MEMORY, AND SOCKETS

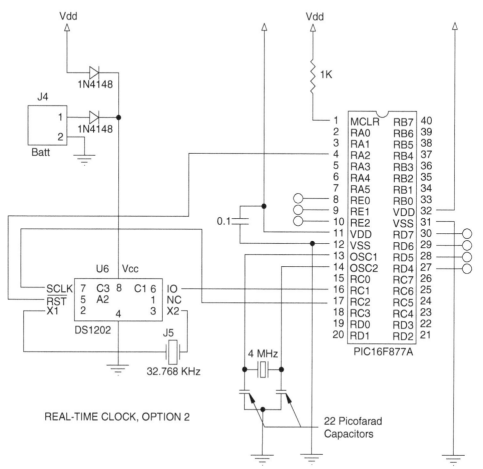

Figure 7.6 Clock implemented using IC DS1202. (To the MCU, this clock IC looks like a set of memory locations.)

Note 2 *There are a total of six empty sockets—U3, U4, U5, U6, U7, and U10—on the LAB-X1 board as received. Though more than one socket can be occupied by an IC at any one time, it is best if only one IC is experimented with at any one time. This will ensure that there are no conflicts between the various devices. If the extreme right RS485 socket U10 is to be used, the RS 232 IC in the socket just to the left of it, in U9, must be removed. One of these two communication chips can remain in place at all times in that the communications circuitry does not conflict with the memory locations.*

THE CLOCK ICS (IN SOCKET U6)

The two 8-pin Dallas Semiconductor clock ICs are interchangeable, and each of them goes into the existing empty socket U6. The DS1302 is the successor to the DS1202.

The NJU6355 also goes into socket U6, but it is not pin-for-pin compatible with the Dallas Semiconductor chips. Fortunately, it too needs to have its crystal between pins 2 and 3, and its other lines can share the connections to the PIC 16F877A.

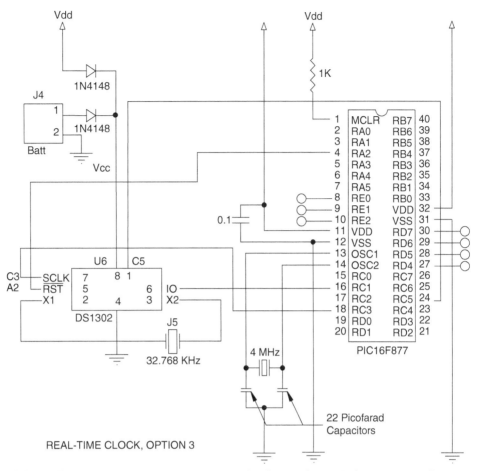

Figure 7.7 Clock implemented using IC DS1302. (To the MCU, this clock IC Looks like a set of memory locations.)

Before you can use the NJU6355, the DS1202, or the DS1302, you have to install a crystal between pins 2 and 3 of the chip socket. This has to be a 32.768 kHz crystal and it is to be installed at jumper J5 next to the real-time clock IC socket. If you do not have a crystal in place, the program will show the date and other items on the LCD, but the clock will not move forward.

If you want to have battery backup for the clock, you need to install a battery at jumper J4 (at the edge of the board next to U10). The pins for this jumper are already on the board when you receive it. The IC will accept from 2.0 to 5.5 volts, so three AA cells in series can provide an inexpensive backup power source. (The power drawn by this IC is 300 nano-amps at 2.0 volts. Two AA cells may not provide enough voltage because of the voltage drop across the in-line diode, so I have recommended three cells.)

The DS1302 (in Socket U6) The DS1302 is the successor to the DS1202. The DS1302 IC is very similar except for backup power capability and seven additional bytes of scratch pad memory. See the datasheet for more specific details.

The emphasis in the program we will develop is to see how we get the data to and from the real-time clock. Setting the clock is going to be done in the program startup routine, and the time cannot be modified once the program is running. If you want that, you can add that to the program you write.

The DS1302 has 31 RAM registers. When you want to send or receive data to the IC, the data can be transferred to and from the clock/RAM one byte at a time, or in a burst of up to 31 bytes.

The LTC1298 12-Bit A-to-D Converter (*Also Used in Socket U6*)

For our purposes, in making instruments and controllers, 8-, 10-, and 12-bit A-to-D converters are used as interfaces between sensors and microprocessors. Sensors usually provide a change in resistance, inductance, or capacitance as some other factor is manipulated. These changes are usually very small and need to be amplified and digitized so they can be manipulated in a digital environment. The interface that converts these small analog signals to useful digital information is the A-to-D converter. Getting comfortable with A-to-D converters is an important part of making instruments and controllers.

MicroEngineering Labs provide a program on their Web site that shows how to read the 12-bit LTC1298. It's shown in Program 7.4.

Program 7.4 Program to read from 12-bit LTC1298 A-to-D chip by microEngineering Labs

```
; PICBASIC PRO program to read LTC1298 ADC by microEngineering
; Labs.
; Define LOADER_USED to allow use of the boot loader.
; This will not affect normal program operation.
DEFINE LOADER_USED 1     ;
DEFINE LCD_DREG PORTD    ; define LCD pins
DEFINE LCD_DBIT 4        ;
DEFINE LCD_RSREG PORTE   ;
DEFINE LCD_RSBIT 0       ;
DEFINE LCD_EREG PORTE    ;
DEFINE LCD_EBIT 1        ;
INCLUDE "MODEDEFS.BAS"   ;
                         ; alias pins
CS VAR PORTC.5           ; chip select
CK VAR PORTC.3           ; clock
DI VAR PORTA.2           ; data in
DO VAR PORTC.1           ; data out
                         ; allocate variables
ADDR VAR BYTE            ; channel address / mode
```

(*Continued*)

Program 7.4 Program to read from 12-bit LTC1298 A-to-D chip by microEngineering Labs (*Continued*)

```
RESULT VAR WORD          ;
X VAR WORD               ;
Y VAR WORD               ;
Z VAR WORD               ;
HIGH CS                  ; chip select inactive
ADCON1 = 7               ; set PORTA, PORTE to digital
LOW PORTE.2              ; LCD R/W line low (W)
PAUSE 100                ; wait for LCD to start
GOTO MAINLOOP            ; skip subroutines
                         ; subroutine to read a/d converter
GETAD:                   ;
  CS = 0                 ; chip select active
                         ; send address / mode -
                         ; start bit, 3 bit addr, null bit]
  SHIFTOUT DI, CK, MSBFIRST, [1\1, ADDR\3, 0\1]
  SHIFTIN DO, CK, MSBPRE, [RESULT\12]; get 12-bit result
  CS = 1                 ; chip select inactive
RETURN                   ;
                         ; subroutine to get x value (channel 0)
GETX:                    ;
  ADDR = %00000101       ; single ended, channel 0, MSBF high
  GOSUB GETAD            ;
  X = RESULT             ;
RETURN                   ;
                         ; subroutine to get y value (channel 1)
GETY:                    ;
  ADDR = %00000111       ; single ended, channel 1, MSBF high
  GOSUB GETAD            ;
  Y = RESULT             ;
RETURN                   ;
                         ; subroutine to get z value
                         ; (differential)
GETZ:                    ;
  ADDR = %00000001       ; diff (ch0 = +, ch1 = -), MSBF high
  GOSUB GETAD            ;
  Z = RESULT             ;
RETURN                   ;
                         ;
MAINLOOP:                ;
  GOSUB GETX             ; get x value
  GOSUB GETY             ; get y value
  GOSUB GETZ             ; get z value
  LCDOUT $FE, 1, "X=", #X, ; send values to LCD
  "Y=", #Y, "Z=", #Z
  PAUSE 100              ; do it about 10 times a second
GOTO MAINLOOP            ; do it forever
END                      ; end program
```

Program 7.4 reads three values from the A-to-D converter and displays them as X, Y, and Z values on the LCD. The 1298 is a two-channel device, and the two signals are read from pins 2 and 3 on the device. The third value being displayed on the LCD is the differential between the two values, meaning that the device is now being used for looking at the two inputs, not as individual inputs but as one signal across the two lines (as compared to two signals between each of the pins and ground).

The two channels are connected to the two pins at J5. These are the two pins that the crystal for the clocks goes across and, as mentioned before, *because of this there is a hardware conflict between using the clock chips and the A-to-D converter.*

The LTC 1298 can provide a maximum of 11.1 thousand samples per second. The device accepts an analog reference voltage between –0.3 and Vcc +0.3 volts, so the signals to be read must be conditioned to reflect these requirements.

Sockets U7 (and U8)

Sockets U7 and U8 are designed for temperature-sensing experiments. (U8 is a three-hole group for soldering in a three-wire temperature-sensing device and is located next to U7.) How this is done is demonstrated in Programs 7.5 and 7.6.

The DS1820 temperature reading device goes in socket U7.

The DS1620 temperature sensor has to be soldered into socket U8.

Program 7.5 Using the DS1820 (Program to read temperature by microEngineering Labs)

```
; This microEngineering Labs program can be found on their
; Web site. Use the latest version.
; PICBASIC PRO program to read DS1820 1-wire temperature sensor
; and display the temperature on the LCD
DEFINE LCD_DREG PORTD     ; define lcd pins
DEFINE LCD_DBIT 4         ;
DEFINE LCD_RSREG PORTE    ;
DEFINE LCD_RSBIT 0        ;
DEFINE LCD_EREG PORTE     ;
DEFINE LCD_EBIT 1         ;
                          ; allocate variables
COMMAND VAR BYTE          ; storage for command
I VAR BYTE                ; storage for loop counter
TEMP VAR WORD             ; storage for temperature
DQ VAR PORTC.0            ; alias DS1820 data pin
DQ_DIR VAR TRISC.0        ; alias DS1820 data direction pin
                          ;
                          ;
```

(Continued)

Program 7.5 Using the DS1820 (Program to read temperature by microEngineering Labs) (*Continued*)

```
ADCON1 =%00000111        ; set PORTA and PORTE to digital
LOW PORTE.2              ; lcd r/w line low (w)
PAUSE 100                ; wait for lcd to start
LCDOUT $FE, 1, "TEMP IN DEGREES C"   ; display sign-on message
                         ;
; mainloop to read the temperature and display on lcd
MAINLOOP:                ;
GOSUB INIT1820           ; init the DS1802
COMMAND = %11001100      ; issue skip rom command
GOSUB WRITE1820          ;
COMMAND = %01000100      ; start temperature conversion
GOSUB WRITE1820          ;
PAUSE 2000               ; wait 2 seconds for conversion to
                         ; complete
GOSUB INIT1820           ; do another init
COMMAND = %11001100      ; issue skip rom command
GOSUB WRITE1820          ;
COMMAND = %10111110      ; read the temperature
GOSUB WRITE1820          ;
GOSUB READ1820           ;
                         ; display the decimal temperature
LCDOUT $FE, 1, DEC (TEMP >> 1), ".", DEC (TEMP.0 * 5),_
" DEGREES C"
GOTO MAINLOOP            ; do it forever
                         ; initialize DS1802 and check for
                         ; presence
INIT1820:                ;
LOW DQ                   ; Set the data pin low to init
PAUSEUS 500              ; wait > 480us
DQ_DIR = 1               ; release data pin (set to input
                         ; for high)
PAUSEUS 100              ; wait > 60us
IF DQ = 1 THEN           ;
LCDOUT $FE, 1, "DS1820 NOT PRESENT"   ;
PAUSE 500                ;
GOTO MAINLOOP            ; try again
ENDIF                    ;
PAUSEUS 400              ; Wait for end of presence pulse
RETURN                   ;
                         ; write "command" byte to the ds1820
WRITE1820:               ;
FOR I = 1 TO 8           ; 8 bits to a byte
  IF COMMAND.0 = 0 THEN  ;
    GOSUB WRITE0         ; write a 0 bit
  ELSE                   ;
    GOSUB WRITE1         ; write a 1 bit
```

(*Continued*)

Program 7.5 Using the DS1820 (Program to read temperature by microEngineering Labs) (*Continued*)

```
      ENDIF                       ;
        COMMAND = COMMAND >> 1    ; shift to next bit
      NEXT I                      ;
    RETURN                        ;
                                  ; write a 0 bit to the DS1802
    WRITE0:                       ;
      LOW DQ                      ;
      PAUSEUS 60                  ; low for > 60us for 0
      DQ_DIR = 1                  ; release data pin (set to input
                                  ; for high)
    RETURN                        ; write a 1 bit to the DS1820
    WRITE1:                       ;
      LOW DQ                      ; low for < 15us for 1
      @NOP                        ; delay 1us at 4mhz
      DQ_DIR = 1                  ; release data pin (set to input
                                  ; for high)
      PAUSEUS 60                  ; use up rest of time slot
    RETURN                        ; read temperature from the DS1820
    READ1820:                     ;
      FOR I = 1 TO 16             ; 16 bits to a word
        TEMP = TEMP >> 1          ; shift down bits
        GOSUB READBIT             ; get the bit to the top of temp
      NEXT I                      ;
    RETURN                        ; read a bit from the DS1820
    READBIT:                      ;
      TEMP.15 = 1                 ; preset read bit to 1
      LOW DQ                      ; start the time slot
      @NOP                        ; delay 1us at 4mhz
      DQ_DIR = 1                  ; release data pin (set to input
                                  ; for high)
      IF DQ = 0 THEN              ;
        TEMP.15 = 0               ; set bit to 0
      ENDIF                       ;
      PAUSEUS 60                  ; wait out rest of time slot
    RETURN                        ;
    END                           ; end
```

Program 7.6 Using the DS1620 (microEngineering Labs program to read temperature)

```
; This microEngineering Labs program, too, can be found on
; their Web site. Use the latest version.
; PICBASIC PRO program to read DS1620 three-wire temperature
; sensor
; and display temperature on the LCD
```

(*Continued*)

Program 7.6 Using the DS1620 (microEngineering Labs program to read temperature) (*Continued*)

```
INCLUDE "MODEDEFS.BAS"      ;
DEFINE LCD_DREG PORTD       ; define lcd pins
DEFINE LCD_DBIT 4           ;
DEFINE LCD_RSREG    PORTE;
DEFINE LCD_RSBIT 0          ;
DEFINE LCD_EREG PORTE       ;
DEFINE LCD_EBIT 1           ;
                            ; alias pins
RST VAR PORTC.0             ; reset pin
DQ VAR PORTC.1              ; data pin
CLK VAR PORTC.3             ; clock pin
                            ; allocate variables
TEMP VAR WORD               ; storage for temperature
LOW RST                     ; reset the device
ADCON1 =%00000111           ; set PORTA and PORTE to digital
LOW PORTE.2                 ; lcd r/w line low (w)
PAUSE 100                   ; wait for lcd to start
LCDOUT $FE, 1, "TEMP IN DEGREES C"   ; display sign-on message
                            ; loop to read the temp and display
                            ; on lcd
MAINLOOP:                   ;
RST = 1                     ; enable device
SHIFTOUT DQ, CLK, LSBFIRST, [$EE]    ; start conversion
RST = 0                     ;
PAUSE 1000                  ; wait 1 second for conversion to
                            ; complete
RST = 1                     ;
SHIFTOUT DQ, CLK, LSBFIRST, [$AA]    ; send read command
SHIFTIN DQ, CLK, LSBPRER, [TEMP\9]   ; read 9 bit temperature
RST = 0                     ;
                            ; display the decimal temperature
LCDOUT $FE, 1, DEC (TEMP >> 1), ".", DEC (TEMP.0 * 5),_
"DEGREES C"
GOTO MAINLOOP               ; do it forever
END                         ; end program
```

8

SERIAL COMMUNICATIONS: SOCKETS U9 AND U10

If all you need is a quick serial communications implementation for your project, Program 8.1 (from the microEngineering Labs Web site) gives all the code you need to read and write to the USART. Combine the single character code in a loop to read and write more than one character (in other words, create strings).

Program 8.1   RS232 Communications (Program to communicate with a computer)

```
; This microEngineering Labs program can be found on their Web
; site. Use the latest version.
CLEAR                    ; read and write hardware usart
DEFINE OSC 4             ; osc speed
B1 VAR BYTE              ; initialize usart
  TRISC = %10111111      ; set TX (PORTC.6) to out, rest in
  SPBRG = 25             ; set baud rate to 2400
  RCSTA = %10010000      ; enable serial port and continuous receive
  TXSTA = %00100000      ; enable transmit and asynchronous mode
                         ;
                         ; echo received characters in infinite loop
LOOP:                    ;
GOSUB CHARIN             ; get a character from serial input, if any
  IF B1 = 0 THEN LOOP    ; no character yet
  GOSUB CHAROUT          ; send character to serial output
GOTO LOOP                ; do it forever
                         ;
CHARIN:                  ; sub to get a character; from usart receiver
B1 = 0                   ; preset to no character received
  IF PIR1.5 = 1 THEN     ; if receive flag then...
    B1 = RCREG           ; get received character to b1
  ENDIF                  ;
```

(*Continued*)

Program 8.1 RS232 Communications (Program to communicate with a computer) (*Continued*)

```
RETURN                        ; go back to caller
                              ; subroutine to send a character
CHAROUT:                      ; to usart transmitter
IF PIR1.4 = 0 THEN CHAROUT    ; wait for transmit register empty
  TXREG = B1                  ; send character to transmit
                              ; register
RETURN                        ; go back to caller
END                           ; end program
```

On the other hand, if you need a greater understanding of what is going on, read on.

The LAB-X1 lets us experiment with two types of serial communications. The board comes with hardware for the RS232 standard on board and an empty socket that can be configured with the RS485 protocol (a line driver IC must be added). Only one type of communications can be active at any one time, and the chip that is not being used must be removed from the board. The RS232 communications are routed to the DB-9 female connector on the board, and there are PC board holes for a *three-pin connector at J10 for RS485* communications. The IC required by the RS485 is the SN175176A or equivalent line driver.

The two standards are similar and, simply stated, using RS485 allows you to go longer distances and the communication is more noise tolerant. (This is related to the capacitance of the lines used and stronger drivers being employed. The discussion of the why and how of this is beyond the scope of this book.)

The compiler supports communications to both standards and the specified compiler commands should be used whenever possible. Writing your own sequences for controlling serial communications is counterproductive, even though it might be instructive. The compiler uses the same commands to access both standards, and the hardware you put in place determines how the signals are sent out and received. (Take a minute to look at Figures 8-1 and 8-2 to see what the complications are.)

Communications are timed according to the specification of the oscillator. For the proper timing to be achieved, the OSC command has to be set to the actual oscillator frequency in use. If the frequencies are not matched, communications will be sped up or slowed down (in speed) based on the extent of the mismatch. If you are actually using a 4 MHz oscillator and specify OSC 20 in your program, the communications will slow down to one-fifth the specified speed because the system is actually running at one-fifth the 20 MHz speed.

In order to experiment with communications, we need to be able to communicate with an external device. The easiest device to use is a personal computer set up with a dumb terminal program. The various versions of Microsoft Office Works software all contain a terminal program that you can access and use.

A dumb terminal may be set up by following these instructions:

Go to the Start menu in your PC command trough, then select

Programs

SERIAL COMMUNICATIONS: SOCKETS U9 AND U10

Accessories

Communications

HyperTerminal

If for some reason HyperTerminal does not show up here, go to HELP under the Start menu, search for "HyperTerminal," and then click "Finding in 2000." This will give you a window with a link to HyperTerminal. Downloads are free.

Set up the HyperTerminal for

8 bits

No parity

I stop bit

2400 baud

This is what the system defaults to with the PICBASIC PRO Compiler. We will use these settings for all our experiments. Set it up and save your terminal to the desktop for easy access.

Connect the serial cable between the computer and the MCU. Since this same cable is also used for programming the MCU, if you have only one serial port, you must disconnect it from the serial programmer after every programming session when using a serial programmer. It's just one more reason for opting for the USB programmer. The wiring for the connector to the serial port is shown in Figure 8.1

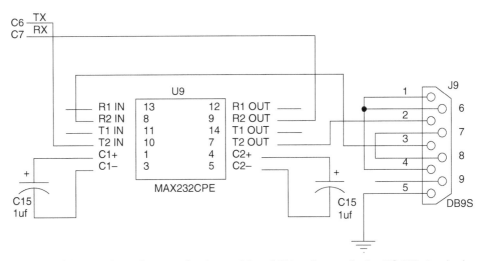

Figure 8.1 RS232 Communications wiring. (Wiring diagram for the RS 232 standard – There are actually two RS232 drivers on the MAX232CPE. The unconnected I/O lines belong to the second driver on this chip.)

When and How Will I Know If It Is Working?

Once set up properly, whatever is sent out by the LAB-X1 will show up on the HyperTerminal screen, and whatever is typed in at the computer keyboard will show up on the LAB-X1 LCD and the HyperTerminal screen.

We will be using the hardware serial output command HSEROUT, which applies to the first COM port on the LAB-X1. (The LAB-X1 has only one port, so HERSOUT2 is not applicable for use with this MCU. The most obvious use is for data collection and conversion with useful filtering where the data comes in on one port [or from an instrument], is translated and filtered, and then goes out on the other port.)

Let's write a simple program (see Program 8.2), with no safety or error correction interlocks, to send a series of 75 uppercase "A"s to the computer one "A" at a time with no delay between transmissions. Seventy-five characters will fit on one line with the carriage return. This will keep the LAB-X1 busy for about 0.25 seconds every time you press the reset pushbutton while adjusting the terminal settings, if necessary.

Program 8.2 RS232 Communications (Program to send information to the computer)

```
CLEAR                       ; Clear the Ram
DEFINE OSC 4                ; Define the oscillator speed
DEFINE HSER_RXSTA 90h       ; ]setting up the communications
DEFINE HSER_TXSTA 20h       ; ]variables and baud rate
DEFINE HSER_BAUD 2400       ; ]see PBP manual for details
DEFINE HSER_SPBRG 25        ; ]
HSEROUT [$D, $A, $A]        ; a carriage return and two line feeds
ALPHA VAR BYTE              ; set counter variable
FOR ALPHA =1 TO 75          ;
   HSEROUT ["A"]            ; loop to send out the 75 'A' characters
NEXT ALPHA                  ;
END                         ; end program
```

For the communications protocol, we will match the settings of the HyperTerminal. As indicated in the compiler manual, this is done with the arguments in the HSEROUT command and by the protocol related DEFINEs in the program. This is what is implemented in Program 8.2. Before you go any further, the preceding program must be made operational.

Next, we need to receive information from the computer and display it on the two lines of the LCD. We will set it up so the LCD will be cleared after every 20 characters received, so we don't run out of space on line 1 of the display. The operant command to receive data is:

```
HSERIN {ParityLabel, }{Timeout, Label,}[Item{,...}]
```

The DEFINEs in the first program segment define the variables being used. "Timeout" and "Label" are optional and are used to allow the program to continue if characters are not received in a timely manner. Timeout is specified in milliseconds. See the more detailed discussion in the compiler manual. In our case, the timeout means that the program will jump to the sending routine whenever there is nothing in the receiver buffer. The receiver buffer has preference as set up. However, you need to keep in mind that the receive buffer is only 2 bytes long, so we cannot linger on the send side too long before checking on the receive buffer again.

Things to keep in mind when receiving information:

- Even though certain control characters do not show up onscreen, they will still be counted as characters unless they are filtered out.
- Some characters may not be implemented at all depending on the character set recognized by the software in the two processors. This means that the filters you design have to be carefully designed to take care of all possibilities.
- The receiving buffer is only two characters long. This is a very important consideration.
- We have not taken any precautions for transmission/reception errors, and so on, and that can get more complicated than we need to cover at this level of our expertise. Meaning that we will assume there are no hardware errors to disturb what we are doing.

We can write a short program (see Program 8.3) to receive and display information on the LAB-X1. Since the information will be displayed on the LCD, we have to include all the usual code for the LCD in our program.

Program 8.3 RS232 Communications (Program to receive and display information from the computer)

```
CLEAR                            ; clear memory
DEFINE OSC 4                     ; osc
DEFINE LCD_DREG PORTD            ; define LCD registers and bits
DEFINE LCD_DBIT 4                ;
DEFINE LCD_RSREG PORTE           ;
DEFINE LCD_RSBIT 0               ;
DEFINE LCD_EREG PORTE            ;
DEFINE LCD_EBIT 1                ;
                                 ;
CHAR VAR BYTE                    ; variables used in the routine
                                 ; storage for serial character
COL VAR BYTE                     ; keypad column
ROW VAR BYTE                     ; keypad row
KEY VAR BYTE                     ; key value
LASTKEY VAR BYTE                 ; last key storage
                                 ;
ADCON1 = %00000111               ; set PORTA and PORTE to digital
```

(Continued)

Program 8.3 RS232 Communications (Program to receive and display information from the computer) (*Continued*)

```
LOW PORTE.2               ; LCD R/W line low (W)
PAUSE 500                 ; wait for LCD to startup
OPTION_REG.7 = 0          ; enable PORTB pullups
                          ;
KEY = 0                   ; initialize vars
LASTKEY = 0
                          ;
LCDOUT $FE, 1             ; initialize and clear display
                          ;
LOOP:                     ;
HSERIN 1, TLABEL, [CHAR]  ; get a char from serial port
LCDOUT CHAR               ; send char to display
                          ;
TLABEL:                   ;
GOSUB GETKEY              ; get a keypress if any
IF (KEY != 0) AND (KEY != LASTKEY) THEN   ;
  HSEROUT [KEY]           ; send key out serial port
ENDIF                     ;
LASTKEY = KEY             ; save last key value
GOTO LOOP                 ; do it all over again
                          ;
GETKEY:                   ; subroutine to get a key from keypad
  KEY = 0                 ; preset to no key
  FOR COL = 0 TO 3
    PORTB = 0             ; all output pins low
    TRISB = (DCD COL) ^ $FF ; set one column pin to output
    ROW = PORTB >> 4      ; read row
    IF ROW != %00001111 THEN GOTKEY ;
  NEXT COL                ;
  RETURN                  ; no key pressed
GOTKEY:                   ; change row and col to ASCII numb
  KEY = (COL * 4) + (NCD (ROW ^ $F)) + "0" ;
RETURN                    ; subroutine over
END                       ; end of program
```

Next, we combine the send and receive programs to give us full communications between the LAB-X1 and the HyperTerminal program in the computer. (This is left to you. At this stage, you should have no problem with implementing this; however, a couple of hints are provided.)

The HyperTerminal software takes care of receiving, displaying, and sending characters, without need for any further modification by us.

The LAB-X1 software must receive characters from the terminal program and display them on the LCD. It also has to read the keyboard and send what it sees to the terminal. The receiving and sending must be in the same main loop.

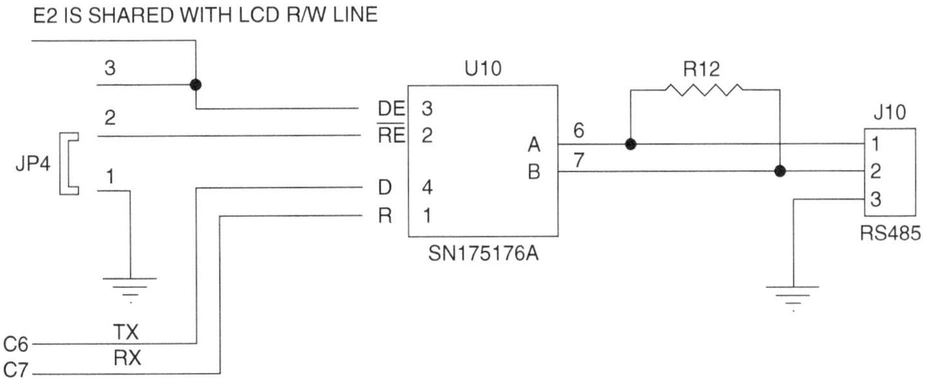

Figure 8.2 RS485 communications wiring. (Wiring diagram for the RS 485 standard.)

Using the RS485 Communications

In order to use the RS485 serial communications standard (see Figure 8.2), pins must be installed in JP4 to enable the ground connections and allow J10 to carry the communications. As mentioned earlier, the RS232 IC in U9 must be removed. After these changes, the operations will be similar to the RS-323C standard.

9

USING LIQUID CRYSTAL DISPLAYS: AN EXTENDED INFORMATION RESOURCE

General

The use of LCD (liquid crystal display) modules is covered in great detail in this chapter because they form an important part of any project based on the use of the PIC line of microprocessors and the PROBASIC PRO compilers. We will consider the popular 2-line-by-16-character display in detail, but the information is applicable to most LCDs on the market today.

The PICBASIC PRO Compiler offers full support for the 2-line-by-20-character display provided on the LAB-X1 board, as well as for other similar displays controlled by the Hitachi HD44780U and compatible controllers. Before a display can be used, it is necessary to tell the compiler where the display is located in memory. This is done by setting the value of a number of DEFINEs that have been named and predefined in the compiler. These DEFINEs let you write to any LCD at any memory location in your project with the compiler (assuming there are no wiring conflicts). The specific DEFINEs related to the control of the LAB-X1 display are as follows:

The DEFINEs described next are for an LCD controlled from PORTD and PORTE, as is the case for the LAB-X1.

Identify the port connected to the LCD data pins:

```
DEFINE     LCD_DREG    PORTD
```

Decide how many bits of data to use, along with the starting bit. This can be a 0 or a 4 for the data starting bit, and 4 or 8 for the number of bits used:

```
DEFINE     LCD_DBIT    0    (or 4)
DEFINE     LCD_BITS    4    (or 8)
```

Specify the register that will contain the register selection bit, and the number of the bit that will be used to select the register:

```
DEFINE      LCD_RSREG    PORTE
DEFINE      LCD_RSBIT    0
```

When we transfer the data, we must enable the transfer by toggling a bit, and the port and bit for doing this are defined with the following two lines:

```
DEFINE      LCD_EREG     PORTE
DEFINE      LCD_EBIT     1
```

Decide whether we are going to read data from, or write data to, the LCD. This is the read/write bit, and this bit is defined with the following two lines:

```
DEFINE      LCD_RWREG    PORTE
DEFINE      LCD_RWBIT    2
```

Most of the time we do not need to read data from the LCD, so this bit can be made and left low:

```
LOW PORTE.2
```

Set LCD R/W pin to low (if write-only is to be implemented). If we are *not ever going to read* from the LCD module, the preceding bit can be set and left low, or it can be tied low with hardware. (Doing it this way will save one control line on the PIC.)

Since the PIC 16F877A has analog capability, like any other similar PIC it will come up in analog mode on startup and reset. The MCU must be changed to digital mode by setting the appropriate A-to-D control register bits before we can use the digital capabilities of the PIC. We need the digital functions of PORTE to control the LCD. This is done with . . .

```
ADCON1 = %00000110
```

This instruction makes all of PORTA and PORTE digital (%00000111 can also be used). Other specifications will provide different results. For a detailed discussion of this, see the datasheet section on A-to-D conversions.

The LCD takes a considerable time to initialize itself after startup, so we have to wait about 500 milliseconds before writing to it. If there are a lot of other tasks that will take place before the first write to the LCD, this time can be reduced. (A trial-and-error approach can be used to determine the minimum time needed.)

```
PAUSE 500    ; Wait .5 secs. for LCD to start up
```

Usually, the first command to the LCD is used to clear the display and write to it on line 1, but I am showing it as two lines, where the first line clears the display and the second line positions the cursor at the first position on the first line and prints the word "Blank."

```
LCDOUT $FE, 1               ; Clear the LCD
LCDOUT $FE, $80, "Blank"    ; written to the 1st line 1st position
                            ; of the LCD
```

All commands (as opposed to characters) sent to the LCD are preceded by the code $FE or decimal 254. Some of the codes are as described in Table 9.1.

TABLE 9.1 LCD CODE LISTINGS

$FE,	$01	Clear Display. (An uncleared display shows dark rectangles in all the spaces, which may appear blotchy and irregular).
$FE,	$02	Go to home. Position one on line one.
$FE,	$0C	All cursors OFF. This is the default condition on startup.
$FE,	$0E	Underline Cursor ON
$FE,	$0F	Underline Cursor OFF. This is the default condition on startup.
$FE,	$10	Mover cursor right one position.
$FE,	$14	Mover cursor left one position.
$FE,	$80	Move cursor to position one of line one.
$FE,	$C0	Move cursor to position one of line two.
$FE,	$94	Move cursor to position one of line three.
$FE,	$D4	Move cursor to position one of line four.

All these and other codes are described in detail in the Hitachi datasheet (see Table 9.1). You must learn where these are located so you can refer to them when necessary.

These commands apply to all LCDs using the Hitachi HD44780U controller or its equivalent. See the datasheet for this controller for more detailed information. This controller has many commands not shown here, including limited graphic capability (within the characters) and the ability to display Japanese kana characters.

It is useful to have the full 40+ page datasheet on hand whenever doing anything more than sending text to the LCD. The Hitachi 44780 datasheet can be downloaded at no charge from the following Web site:

- www.alldatasheet.com/datasheet-pdf/pdf/63673/HITACHI/HD44780.html

LCD displays require that each line be addressed with its own starting position as indicated earlier. The exception is that most 16-character single-line displays are designed such that the first eight characters start at $80, and the next eight start at $C0, so that the 16 characters appear to be on two lines to the controller though they are displayed as one line on the LCD module. Also note that lines 3 and 4 of four-line displays also have an irregular addressing scheme.

If more characters than can be displayed on a line are sent to the LCD, they will be stored in the memory in the LCD and can be scrolled across the screen when needed. The number of characters that an LCD can store in its display memory is a property of

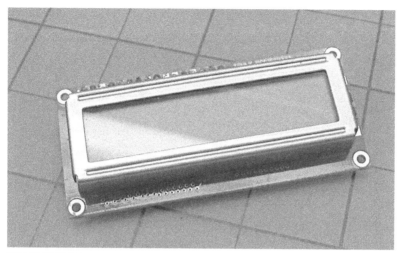

Figure 9.1 A 2-line-by-16 character LCD module. (One- or two-line LCDs can add a tremendous amount of utility to a project.)

the LCD as determined by the manufacturer. You can also scroll the display up and down if you design the commands, and write the software, to do so. This is not built into the LCD or the controller software. A typical 16 × 2 display is shown in Figure 9.1

Using LCDs in Your Projects

It is generally agreed that most projects benefit from having a one- or two-line display incorporated in them. However, these displays tend to be rather pricey (about $50) when provided with the necessary controlling IC, and quite reasonable (about $5 to $6) when bought without the supporting package. Since a PIC microcontroller can be purchased for about $5, we should be able to have a complete display unit for a marginal cost of around $10 if we can figure out how to program our PIC microcontroller to control the display.

The readily available and inexpensive 2-line-by-16-character LCD ($6 at All Electronics) offers us the ability to display information in a limited but useful way in our projects. Mastering the use of this LCD display means we have gained the expertise to write any character, at any location, at any time, or in response to any event, whenever we want. This chapter addresses this problem in detail and tells you what you need to know and do to make these displays inexpensive additions to all of your projects.

In this chapter, you will learn how to control an LCD. The code you create will be able to be incorporated into almost any PICBASIC PRO program and will control the LCD from any available half port (nibble) and three other free I/O lines. The code will be more linear than it needs to be so you can see exactly what is going on, but once you understand what needs to be done, you can write more compact and sophisticated code that will get the job done the way you want.

Understanding the Hardware and Software Interaction

The hardware we are considering consists of an LCD with an integral controller that is incorporated in the display by the display manufacturer. In our particular case, this is the Hitachi HD44780U controller. Displays are available without this or any other controller, but controlling displays without a controller is way beyond the scope of this book. For our projects, be sure you buy only those units that have this controller built in as a part of the display. Most do.

Controlling the display consists of telling this controller what we want it to do. The instructions are easy to learn and allow you to control each and every pixel and all the functions that the display can perform, with relative ease. You do not have to read or understand the rather extensive 40+ page datasheet that Hitachi provides for this controller, but it is well worth the trouble to download the data sheet and study it. You do not *need* to download this file either, but you should know where to find it if you need it. We will go over almost everything you need to know to control the display as a part of this exercise. The exercises at the end of the chapter will take it to the next level.

In the LCD display, the imbedded controller provides the interface between the user and the display. The controller is the Hitachi HD44780U. This very powerful controller gives you complete control over the LCD. It allows you to address each and every pixel on the display. It also has a built-in set of ASCII characters for use by the user. Our task is to learn how to use this controller to put what we want, when we want, where we want, in the display.

The other common controller is the Epson SED series controller, whose operation and instruction set is very similar to that of the Hitachi controller. We will *not* consider the EPSON controller in this book If you do decide to buy a display with this controller to save a few bucks, you should be able to use it after only a small amount of study.

You will find that almost all the smaller liquid crystal displays on the market are controlled by the Hitachi controller mentioned earlier. This means that once you learn to control one display, you can control most of them with the code you create. As a matter of fact, we will write the code in a way that will be universal in its application in that we will define variables like the number of character spaces in the display and the number of lines in the display as part of the program setup. (It is also useful and usually much easier to use the DEFINEs created by the PBP compiler to control the display. We will learn how to do this in a later exercise.)

The addresses of the local memory locations (meaning the ones in the LCD) used by the LCD have already been fixed, as has the instruction set we use to write to the LCD, so we do not have to create any of this rather sophisticated code.

You do not need the full datasheet, which is about 40 pages long, but you do need to know the basic command set that controls the data transfer to your particular display. This is usually provided by the organization you buy the LCD from and consists of two or three pages. You will need to refer to the datasheet only if you want to create special characters, or if you want to display bargraphs and the like on the display. The control the Hitachi controller provides is very comprehensive, but you don't need to be familiar with it to use a display effectively. Everything you need to know will be covered in this exercise, but that should not keep you from learning as much as you can about controlling the displays.

Talking to the LCD

The preceding control codes allow you to configure the display, set display parameters, set the shape and position of the cursor, and so on. To differentiate them from the character commands, each control code must be preceded by a *hexadecimal FE or a decimal 254* to tell the controller that the next character sent to the display will be a control code. After receiving one control code and its argument, the Hitachi controller resets to the data mode automatically.

The controller supports the ASCII standard. All uppercase and lowercase characters and numerals are supported, as are punctuation marks and the standard text support characters. (The controller also supports the display of a set of Japanese kana characters.)

It is also possible to design your own font for use with the displays (though five by seven [or even ten] dots and two lines does limit what can be done). All the information needed to do so is contained in the Hitachi HD44780U datasheets. Greek characters and certain scientific notations would be useful for most scientific applications.

The Hardware

Let's take a closer look at the LCD hardware.

Study the datasheet that came with the LCD. Find the pin-out descriptions and study them. The 16 pins are usually identified as shown in Table 9.2.

Looking at the datasheet provided with the 2-line-by-16-character displays, we find that the control implementation can take place if we have both a port and a few lines available to control the LCD. It does not have to be controlled from any predefined lines. We can select all the lines needed to support the display in our project and they can be on any port we have available. The only requirement seems to be that the four/eight data line be either the contiguous top, or the bottom half, of a port. This is not a particularly demanding requirement other than that it means the smaller PICs cannot be used if we will require many I/O lines in our project. The other three lines needed can be on any of the other ports, and do not all need to be on the same port. Since we are considering

TABLE 9.2 LCD CODE LISTING PIN-OUT IDENTIFICATION OF THE LCD PINS FROM THE DATASHEET

PIN NO.	SYMBOL		
1	VSS	Logic ground	
2	VDD	Logic power 5 volts	
3	VO	Contrast of the display, can usually be grounded.	
4	RS	Register select	) These are
5	R/W	Read/Write	) the 3 control
6	E	Enable	) lines
7	DB0	)	
8	DB1	)	
9	DB2	) This is	
10	DB3	) one port or	
11	DB4	) 8 lines	) half of the port
12	DB5	) of data	) can also
13	DB6	)	) be used
14	DB7	)	) see PBP manual
15	BL	Backlight power	)These two lines can
16	BL	Backlight ground	)usually be ignored

only one PIC in this book, in our case this means we will use the 16F877A. I have included the circuitry and code for this in the Program 9.1 and Figure 9.1 so you can see exactly what needs to be done.

To keep it simple, let's use PORTA and PORTB, because these ports are available on even the smaller MCUs, and if we use a dedicated MCU to control our LCD, it will in all probability be a smaller more inexpensive 18- or 20-pin MCU. Let's use the lines as follows:

Lines 1 to 3 of PORTA as the control lines and

Lines 0 to 7 of PORTB as the data lines

We can redefine these to be more rational addresses whenever we want, and none of the programming will have to change. Just DEFINE what ports and lines you want to use at the top of the program, and the aliases assigned to them will identify them as needed.

With this in mind, let's create the software to control a 2-line-by-16-character display. Once we are happy with what we have created, we can migrate the code to other microcontrollers.

Note *We could also use a one-line (about $5) display, but that would inhibit learning about going to line 2, scrolling the display up and down, and so forth. To experiment with these features, we need to have a display with at least two lines.*

Setting Out Our Design Intent

- Control the 16 × 2 display with a PIC 16F877A microcontroller (the code for the smaller MCUs will be the same).
- Design the software so it can be an integral part of the software for any project.
- Use standard control codes so the project is a virtual plug-in replacement for other displays and in other PICs. (Only minor modifications if any will be required.)
- Use a minimum number of external components, allowing this software project to move between PIC controllers of all descriptions. All we want to do is include the code in our project and connect the display to the selected ports.
- Use the project's regulated 5-volt power supply for everything.

Note *The PIC 16F877A has 33 I/O lines, the display will use 7 of them, so we will have 26 lines left over for our project. Since we don't need all these lines, we could have used the 16F84A. No program changes should be needed, other than changing the line and port addresses in the DEFINES, when we move to the PIC 16F84A.*

HARDWARE NEEDED

We will need the following parts for this project:

- Experimental solderless breadboard
- PIC microcontroller: 16F877A or 16F84A
- One bare 2-line-by-16 character display module (with a Hitachi controller)
- One 4 MHz crystal
- Two 22 pf capacitors
- Regulated 5-volt power supply from the breadboard
- One 470-ohm 0.25-watt resistor
- Some 22-gauge insulated single-strand hookup wire
- 1-kohm 0.12-watt pull-up resistor

PROGRAMMER NEEDED

The programmer needed should be either the microEngineering serial programmer, a parallel programmer, or the new USB programmer.

SOFTWARE NEEDED

The software needed includes the PICBASIC PRO Compiler and its manual.

INFORMATION NEEDED

Information needed includes:

- LCD datasheets that came with the LCD
- The PIC 16F877A datasheet (or the datasheet for the 16F84)

We will go through the software a step at a time, and when done, it will be your job to clean up the software and speed up its operation—for example, to optimize it for the microcontroller you will use in your project(s).

We must pick a specific display to work with so we can develop real working software for it. The display I picked sells for $6 or so and is available with a datasheet from All Electronics. AZ Displays also sells one that seems to be identical—called Model ACM 1602K. The short form datasheets are similar, but the AZ one is in a crisp PDF format and can be downloaded from their Web site for free. Doing so will mean you can have this information open in a window on your computer.

First, we need to investigate how many pins we will need on our PIC microcontroller to interact with our display. This is summarized in Table 9.3.

TABLE 9.3 LCD PIN OUTS (PIN OUT IDENTIFICATION OF THE LCD PINS FROM THE DATASHEET)

PIN NO.	SYMBOL	LEVEL	DESCRIPTION AND NOTES
1	VSS	0V	Ground for logic supply
2	VDD	5.0V	Logic power supply, regulated
3	VO	—	LCD contrast. Can be grounded
4	RS	H/L	H: Data code L: Instruction code
5	R/W	H/L	H: Read mode L: Write mode. Can be tied low in hardware
6	E	H, H>L	Enable signal, pulsed from H to L. Hold at H.
7	DB0	H/L	Data bit 0
8	DB1	H/L	Data bit 1
9	DB2	H/L	Data bit 2
10	DB3	H/L	Data bit 3
11	DB4	H/L	Data bit 4
12	DB5	H/L	Data bit 5
13	DB6	H/L	Data bit 6
14	DB7	H/L	Data bit 7
15	BL	5.0	Plus 5V. Power for back lighting the display.
16	BL	Gnd	Ground for back lighting power.

TABLE 9.4 LCD PIN OUTS (LCD PINS NOT CONTROLLED BY THE MICROPROCESSOR ARE MARKED OUT)

PIN NO.	SYMBOL	PIC PIN	DESIGNATION	
~~1~~	~~VSS~~		~~Not connected to the PIC~~	
~~2~~	~~VDD~~		~~Not connected to the PIC~~	
~~3~~	~~VO~~		~~Not connected to the PIC~~	
4	RS	1	RA2 PortA	
5	R/W	2	RA3 "	
6	E	18	RA1 "	
7	DB0	6	RB0 PortB	
8	DB1	7	RB1 "	
9	DB2	8	RB2 "	
10	DB3	9	RB3 "	
11	DB4	10	RB4 "	)
12	DB5	11	RB5 "	) Half the port can also be used
13	DB6	12	RB6 "	) See PBP manual
14	DB7	13	RB7 "	)
~~15~~	~~BL~~		~~Not connected to the PIC~~	
~~16~~	~~BL~~		~~Not connected to the PIC~~	

Table 9.4 indicates that our microcontroller does not need to be connected to lines 1, 2, 3, 15, and 16 in that these have to do with power connections and not with data I/O. In the preceding case, we will use an 8-bit data buss, and the connection to the PIC 16F877A will be as shown in Figure 9.1.

We can get by with 11 lines, and possibly even 10 if we decide to do without the ability to read from the display memory. This is not usually the case, however, because there are times when we need to set this line high for reading the LCD's busy flag in order to minimize the time used by LCD routines. Nevertheless, we can add about a 20-milliseconds delay to take care of the busy time. We also need to be able to read the display memory if we want to scroll the display up and have access to what is on display lines one and two. Since this would be true for all applications, we have to stay with the 11 lines for 8-bit control. (We would not have to read the display if we kept track of what we had put in the display somewhere else in the memory.)

The eight data lines form a convenient byte, and we can assign one of the ports not being used for anything else. This leaves three lines:

- The E line, which needs to be toggled to transfer data to the LCD
- The RS line, which selects the register
- The (R/W), which sets the read/write status of the operations

Using eight lines for data allows us to generate all the codes and all the characters that the chip has in its memory, but more importantly, it allows the data transfer to be performed in one step. We can also use a 4-bit protocol and transfer half a byte at a time. Using four lines for our control scheme means the LCD can be controlled from just one port (seven lines will be used, leaving us one line to spare). The datasheets tell us that whether we use four lines or eight, they all must be part of one port, and if we are using four lines, they must be the contiguous four high, or the contiguous four low, bits of a port—meaning we cannot use just any random lines for the data buss. The data transfer for the 4-bit protocol must use the *4 high bits on the LCD*, and we must send the data *from the PIC to the display with the high data nibble first and the low data nibble last*. This little gem *is not spelled out in the instructions*, but it's what must be done.

The datasheet also tells us that the display initializes itself on power-up. We can reinitialize it under our control, but it is done automatically on startup and we cannot inhibit that. All we have to do is "not do anything" for about half a second for the self-initialization to complete. The busy flag is set high during startup and initialization, but is indeterminate immediately after initialization starts and for 16.4 milliseconds after the supply voltage reaches 4.5 volts, so we cannot determine how long we have to wait after powering up to start doing what we want. We will set a 0.5-second wait/pause in our program at startup. If that is not long enough, we will come back and increase the waiting time. Wait time is a must. If you do not wait, the system will not start up properly.

Automatic initialization sets the following conditions for the display:

- Display cleared
- Set for 8 bit interface
- Set for 1 line of display
- Set for 5×7 dot matrix display
- Display is turned off
- Cursor is turned off
- Blink is turned off
- Increment between characters is set to 1 (cursor moves over one space automatically)
- Shift is off

The preceding may not be exactly what we want for our purposes, so we will go through an initialization sequence as specified by the instructions. We do not have to go through all the steps, *but we will*, so we have a complete record of what needs to be done for future projects.

The instructions tell us that the following six instructions must be sent to the display during an initialization sequence. The first three instructions are identical but require different waits after each is sent to the display.

150 USING LIQUID CRYSTAL DISPLAYS: AN EXTENDED INFORMATION RESOURCE

TABLE 9.5 LCD STARTUP (SEQUENCE STEPS AND TIMING DELAYS)

STEP	RS LINE	R/W LINE	DATA BYTE	ENABLE LINE
1	1 high	0 low	0011xxxx	Toggle H to L
1A	wait for 4.1 msec			
2	1 high	0 low	0011xxxx	Toggle H to L
2A	wait for 100 µsec			
3	1 high	0 low	0011xxxx	Toggle H to L
3A	wait for 1 msec			
4	1 high	0 low	001110xx	Toggle H to L
5	1 high	0 low	00000001	Toggle H to L
6	1 high	0 low	00000110	Toggle H to L

These instructions are commands rather than data, so the RS (register select) line must be held high while we initialize. See Table 9.5.

The six lines of code in Table 9.5 are explained in detail as follows:

```
                  "x" in a bit = don't care
00110000          ; code to initialize the LCD (this is entered
                  ; 3 times, 1st time)
                  ; load for a command function
                  ; wait at least 4.1ms
                  ;
00110000          ; code to initialize the LCD       (2nd time)
                  ; load for a command function
                  ; wait at least 100us
                  ;
00110000          ; code to initialize the LCD       (3rd time)
                  ; load for a command function
                  ; wait at least 1 millisecond
                  ;
00111000          ; put in 8 bit mode, 2 line, 5X7 dots
                  ; 0
                  ; 0
                  ; 1 =req'd
                  ; 1 =8 bit data transfer
                  ; 1 =2 lines of display
                  ; 0 =5x7 display
                  ; x
                  ; x
                  ; load for a command function
                  ;
```

```
00010100       ; set cursor shift etc
               ; 0
               ; 0
               ; 0
               ; 1 =req'd
               ; 0 =cursor shift off
               ; 1 =shift to right, or left (0)
               ; x
               ; x
               ; load for a command function
               ;
00001111       ; LCD display status, cursor, blink etc
               ; 0
               ; 0
               ; 0
               ; 0
               ; 1 =req'd
               ; 1 =display on
               ; 1 =cursor on so we can see it
               ; 1 =blink on so we can see it
               ; load for a command function
               ;
00000110       ; lcd entry mode set, increment, shift etc
               ; 0
               ; 0
               ; 0
               ; 0
               ; 1 =req'd
               ; 1 =increment cursor in positive dir
               ; 0 =display not shifted
               ; load for a command function
```

At the end of these instructions, the display will have been initialized the way we want it.

There is also this business about the busy flag that we need to be thinking about. The display takes time to do whatever we ask it to do, and the time varies with what we asked it to do. We can wait a few milliseconds between instructions to make sure it has had enough time for the task to complete, or we can monitor the busy flag and as soon as it is not busy we can send the next instruction. Since time is always at a premium, and we want to run as fast as we can, it means we must consider monitoring the busy flag. (This is important because addressing the LCD is one of the most time-consuming parts of most programs, and these small processors are running at only some 20 MHz.)

THE BUSY FLAG

The instruction sheet tells us that the busy flag is bit 5 at location 11100011 in the LCD. Meta Code for waiting for the busy bit in the LCD to clear is as follows:

Busycheck:

Read busybyte

Isolate busybit

If it is busy then goto Busycheck

Return

We isolate the bit and if it is not low, we read the flag byte again. We do this again and again till the bit goes low. As soon as it does, we can write to the LCD and go on with the program.

We also have an interest in having our display be compatible with code generated by the PICBASIC PRO Compiler. Looking at the instructions for the LCDOUT command, we find that the compiler would prefer that the hardware be set up for the following conditions:

- Four data bits DB4 to DB7 connected to PORTA.0 to PORTA.3
- Chip enable at PORTA.3
- Register select at PORTA.4
- Two lines of display are assumed, but we should specify that

If we cannot meet the preceding requirement, we must set the addresses out as DEFINEs in each and every program we write (or we can use an "INCLUDE" statement that includes a program that does this for us). It's only a few lines of code, but we will have to add the code every time, and it compromises compatibility with other systems that will no doubt be set up to meet the compiler standard. It may turn out that the microcontroller you choose will need to have this done in any case, but if it can be avoided, it should be done at this stage of our learning process.

The next thing we need to decide is whether we are going to use the software as an integral part of a program that is running on a larger microcontroller where we can use all ten address lines for the display, or do we want the software to run on a smaller dedicated microcontroller that will need only one serial line to control the display but will have to be added to the total project as a part of hardware we design. For now, let's agree that we will go with a dedicated controller just to run the display. The software for running on a larger controller will be a subset of what we develop so no work is lost here.

The task on the input side is to design the software that will take the serial information received on one pin and output it as 4-bit characters to the LCD with the select, read/write, and enable lines. The work needed to do read the data in is done by the compiler with the SERIN instruction.

Program 9.1 does just this, and is provided by microEngineering Labs on their Web site.

Program 9.1 For a PIC 16F84A (Program to simulate back pack [by microEngineering Labs])

```
; PicBasic Pro program to simulate an LCD Backpack
; Define LCD registers and bits
DEFINE LCD_DREG PORTD     ;
DEFINE LCD_DBIT 4         ;
DEFINE LCD_RSREG PORTE    ;
DEFINE LCD_RSBIT 0        ;
DEFINE LCD_EREG PORTE     ;
DEFINE LCD_EBIT 1         ;
                          ;
CHAR VAR BYTE             ; Storage for serial character
MODE VAR BYTE             ; Storage for serial mode
RCV VAR PORTB.7           ; Serial receive pin
BAUD VAR PORTA.0          ; Baud rate pin - 0 = 2400, 1 = 9600
STATE VAR PORTA.1         ; Inverted or true serial data - 1 = true
                          ;
ADCON1 = %00000111        ; Set PORTA and PORTE to digital
LOW PORTE.2               ; LCD R/W line low (W)
PAUSE 500                 ; Wait for LCD to startup
                          ;
MODE = 0                  ; Set mode
IF (BAUD == 1) THEN       ;
  MODE = 2                ; Set baud rate
ENDIF                     ;
                          ;
IF (STATE == 0) THEN      ;
  MODE = MODE + 4         ; Set inverted or true
ENDIF                     ;
                          ;
LCDOUT $FE, 1             ; Initialize and clear display
                          ;
LOOP:                     ;
  SERIN RCV, MODE, CHAR   ; Get a char from serial input
  LCDOUT CHAR             ; Send char to display
GOTO LOOP                 ; Do it all over again
END                       ; end program
```

This program is for the 16F84A, but it can be used on the 16F877A with appropriate DEFINEs. You have set these DEFINEs many times before now, so this should not be a problem.

The preceding program is for a 16F84A PIC. If you load this program into the PIC, you can connect the 16F84A to the LCD, and any serial information that comes in on PORTB.7 will be displayed on the LCD. Now you can control the LCD from one line on the main processor. (The selected pin does not have to be PORTB.7. Any free pin can be specified as the input data pin in the program.)

The wiring diagram for the 16F84A is shown in Figure 9.2.

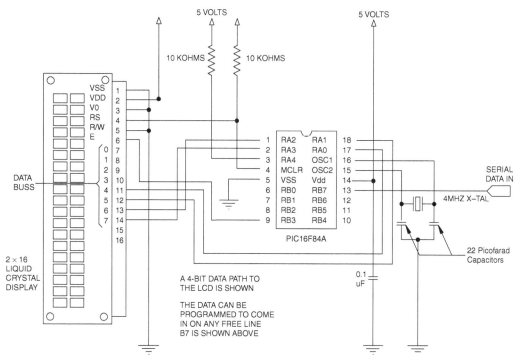

Figure 9.2 Wiring diagram: LCD backpack using a PIC 16F84. (The data can be programmed to come in at any free line. It does not have to be on pin B7. You set the line you want to use in Program 9.1.)

Liquid Crystal Display Exercises

These liquid crystal exercises are to be performed on the LAB-X1 board.

1. Write a program to put the 26 letters of the alphabet and the ten numerals in the 40 spaces that are available on line 1. Insert four spaces between the numbers and the alphabet to fill in the four remaining spaces. Once all the characters have been entered, scroll the 40 characters back and forth endlessly though the 20 spaces visible on line 1.
2. Write a program to bubble the 26 capital letters of the alphabet through the numbers 0 to 9 on line 2 of the LCD. (This means: First, put the numbers on line 2. Then, "A" takes the place of the "0" and all the numbers move over. Then, the "A" takes the place of the "1" and the "0" moves back to position 1 and so on till it gets past the 9. Then, the "B" starts it way across the numbers, and so forth.)
3. Write a program to write the numbers 0 to 9 upside down on line 1. Wait 1 second and then flip the numbers right side up one by one. Provide a time delay between changes. Loop.

4. Write a program to identify the button pressed on the button pad by displaying its row number on line 1 and its column on line 2. Identify each line so you know what is being displayed where. Scroll the two lines up every time a button is pressed. Add delays in the scroll so you can actually see the scrolling take place.

The full instruction table for the LCD is shown in Table 9.6

TABLE 9.6 THE LCD CODE TABLE

COMMAND	R/S	R/W	7	6	5	4	3	2	1	0	DESCRIPTION Fosc=250 KHz	EXECUTING TIME
Clear Display	0	0	0	0	0	0	0	0	0	1	Clears display & returns to address 0	1.64 msec
Cursor at Home	0	0	0	0	0	0	0	0	1	x	Returns cursor to address 0. Also returns the display being shifted to the original position. DDRAM contents remain unchanged	1.64 msec
Entry Mode Set	0	0	0	0	0	0	0	1	I/D	S	I/D: Set cursor moving direction I/D=1: Increment I/D=0: Decrement S: Specify shift of display S=1: The display is shifted S=0: The display is not shifted	40 µsec
Display ON/OFF Control	0	0	0	0	0	0	1	D	C	B	Display D=1: Display on D=0: Display off Cursor C=1: Cursor on C=0: Cursor off Blink B=1: Blink on B=0: Blink off	40 µsec
Cursor / Display Shift	0	0	0	0	0	1	S/C	R/L	x	x	Moves cursor or shifts the display w/o changing DD RAM contents S/C=0: Cursor shift (RAM unchanged) S/C=1: Display shift (RAM unchanged) R/L=1: Shift to the right R/L=0: Shift to the left	40 µsec

(Continued)

COMMAND	RS	RW	7	6	5	4	3	2	1	0	DESCRIPTION Fosc=250 KHz	EXECUTING TIME
Function Set	0	0	0	0	1	DL	N	F	x	x	Sets data buss length (DL), # of display lines (N), and character fonts (F) DL=1: 8 bits F=0: 5x7 dots DL=0: 4 bits F=1: 5x10 dots N=0: 1 line display N=1: 2 lines display	40 μsec
Set CG RAM Address	0	0	0	1	Character generator (CG) RAM address					Sets CG RAM address CGRAM data is sent and received after this instruction	40 μsec
Set DD RAM Address	0	0	1	Display data (DD) RAM address/cursor address						Sets DD RAM address. DD Ram data is sent and received after this instruction	40 μsec
Busy Flag/ Address Read	0	1	BF	Address counter used for both DD & CG RAM						Reads busy flag (BF) and address counter contents	40 μsec
Write Data	1	0	Write data									Writes data into DDRAM or CGRAM	46 μsec
Read Data	1	1	Read data									Reads data from DDRAM or CGRAM	46 μsec

TABLE 9.6 THE LCD CODE TABLE (*CONTINUED*)

Table 9.6 is for the 16-character-by-2-line LCD display sold by "Images Inc. SI" of New York for about $10 plus postage, and is the module used in all the preceding experiments and those shown in the related diagrams. All Electronics has one for $6.

Almost all LCDs with back packs use the preceding scheme.

Part II

THE PROJECTS

10

USING SENSORS (TRANSDUCERS)

General

Before you can make a decision about modifying an existing condition, you have to sense what the current condition is. An awful lot can be done with the five senses we all have, but for many other things that are important in our everyday lives we need help. We get this help from electronic sensors. These sensors convert the information in the system into information we can absorb using our five senses. The preferred sensory input in humans is visual and auditory, but we often also use touch. Smell and taste are used less often and are usually reserved for organic inputs, sometimes those indicating a hazard or danger.

The following are some unusual examples of how we use our five senses.

- **Sight** Automobile traffic is controlled basically by three colored lights.
- **Hearing** The elevator beckons us with a ping.
- **Touch** The cell phone vibrates (if set to do so).
- **Smell** Gasoline contains an additive to help us detect its dangerous presence.
- **Taste** The DEA agent can recognize cocaine by its taste (at least on TV).

When we think about sensing something our human senses cannot detect, we are usually thinking about signals that, to us, are weak or hard to differentiate in the ambient noise. These are signals that need to be filtered and amplified and then converted to signals we *can* recognize. As you read this, a few hundred television and radio broadcasts are zipping past your brain that you are completely oblivious to. If you know what frequencies to look for, how to filter out the signal you want from them, and then how to present it to the human eye and ear, you could watch and hear the data being broadcast. Cable network subscriptions seem to indicate this is worth a lot of money every month!

Thus, there are four main aspects of sensing.

- We must have a knowledge and understanding of the existence of the signal.
- We must find a filter that will isolate and identify the signal.

- We must amplify the signal if it is weak.
- We must convert the signal to accommodate a human sense, or transform it into something detectable by a machine or instrument, which we can then interact with.

Only then are we in a position to make an intelligent decision about the signal.

One more important thing: We need to have a very good idea of what we are trying to do in order to make sure we end up with the signal we are interested in. It won't do to have an instrument that tells us the humidity, when what we were really interested in was the temperature.

The world is full of many exotic and interesting things to experiment with and detect, but in these discussions we will concentrate on those things we might find in an everyday engineering school laboratory or amateur engineer's workshop. We will not discuss anything for which a sensor itself cannot be purchased for less than $30 (U.S. dollars). We will stick to using only a few sensors and employ them in many different ways. The sensors/detectors/transducers we select must be readily available and be *interface-able* with a small microcontroller. Even so, other efforts must be undertaken before we have a viable instrument or controller—something which we will discuss later in Chapters 15 to 22.

We cannot do anything about the signal, however, if we don't know where to find it or know that it even exists. The signal also must have a sensor we can use, that will respond to it, and it is preferable that the sensor respond only to the signal we are interested in. All other responses detract from the task at hand and can be considered part of the general "noise."

Some of the most exquisitely sensitive sensors are biological. The antennae of the moth can recognize the pheromones it has evolved to detect in the parts per billion or less. These sensors are so sensitive they can even detect the gradient in the scent, allowing the moth to move towards its target. Here it is important to note that in a three-dimensional space, one part in 1,000,000,000 means that the molecules being detected are 1000 molecules apart. Since the antennae are many millions of molecules long, quite a few particles are possibly being intercepted at any one time at these seemingly minute concentrations. (We are just beginning to see the use of biological/chemical elements in our most sophisticated and cutting edge electric/electronic instruments.)

In most cases, we are interested in linear space rather than three-dimensional space, as mentioned earlier. In most cases, if we can detect about 1 part in 10,000, we are quite happy. One part in 10,000 refers to being able to detect a change of 1 part within the entire range from 1 to 10,000. To use a real-world example, consider the common voltmeter, which can detect from 0 to 1000 volts with a sensitivity of 0.1 volts. This is the approximate range of our most common instruments, but of course we use both much coarser and much finer instruments also. The following lists some common examples we are all familiar with.

Oven thermometer	0 to 600 degrees	15 degrees
Oral thermometer	95 to 105 degrees	0.1 degrees
House thermostat	32 to 132 degrees	1 degree
Car speedometer	0 to 120 mph	2 mph

Car odometer	0 to 999,000 miles	0.1 miles for short distances
6-inch ruler	0 to 6 inches	1/32 inch at best
Steel tape	0 to 96 inches	1/16 inch
Mechanical stop watch	0 to 30 minutes	0.05 seconds

Since we are interested in the use of sensors that connect to microcontrollers, we will restrict ourselves to those sensors that provide an electrical signal or a signal that is easily converted into an electrical signal. Signals that are easy to interface to a microcontroller are signals that have (changing) electrical properties like...

- Resistance
- Voltage
- Capacitance
- Reluctance
- Frequency

What competencies must be mastered in order to create microcontroller-based instruments and controllers? Basically, you have a signal that must be transformed into a reading of some sort in a one- or two-line display, or something similar. What you must master is the process between the two. Thus, the tasks involved break down this way:

1. Understanding the problem (this is much more important than it seems)
2. Capturing the signal
3. Filtering out the part you want
4. Conditioning the signal to make it acceptable to the PIC
5. Converting it to a digital format
6. Manipulating the digital data to create a readable value
7. Doing one or more of the following:
    - Displaying the value on the LCD
    - Transmitting the value to a computer for collection and storage
    - Turning external device(s) on and off as necessary

Essentially, the chapters in this tutorial are devoted to these tasks.

The Most Basic Question We Must Answer Is...

Why would I want to build an instrument when instruments to measure almost everything I am interested in are already available off the shelf?

The answer to this question is neither simple nor short.

First: Almost all the instruments you need *are not available off the shelf*. More accurately stated, only bits and parts of the instruments you require are available. The instruments we will build will be more useful than their generic off-the-shelf siblings, and in some ways are more specifically targeted to the task at hand. Since we know exactly what we need, we will design an instrument that provides exactly what we want. We do not have to compromise on any property of the instrument. Also, our instruments will be able to provide other intelligence functions, like turning other laboratory equipment on and off as needed by our experiments, and as determined by the conditions the instrument is monitoring in real time. This is a very useful feature almost never found on an industrial instrument, but that is absolutely essential if we want to automate our processes.

Second: We will be able to automatically send the information being gathered to a computer for analysis, either in real time or on a deferred basis depending on what our overall needs are. We can also gather a lot more information over a longer period of time with our custom instruments because we will now have the ability to automate the process. Transient phenomena that require constant monitoring over long periods of time and produce only in a few important instances can now be monitored continuously and intelligently without concern or added expense.

Third: Our instruments will be able to make intelligent decisions in real time. If data points that are unexpected or extraordinary are encountered, the instrument can call this to our attention so remedial or special (even human) attention can be given to the problem.

Fourth: On the output side, the ability to turn pumps, fans, heaters, and the like on and off automatically, based on the information sensed by the instrument cannot be dismissed out of hand. Very few off-the-shelf volt-/ohmmeters can turn an ancillary piece of equipment on or off at a given voltage. Most do not even have an output that we could connect to if we so desired. However, with our custom-designed instruments, it will be easy. Few ohmmeters can send the value read to a computer every second or every hour. But with the instruments we will build, it will be easy. No off-the-shelf instruments can be reprogrammed in BASIC with one click of the mouse. With the instruments that we will create, however, this will be the case. The instruments we shall design and build will be intimately familiar to us, so modifying them will be relatively easy. Once the input-output appurtenances have been decided on and connected to the microcontroller, the rest will be controlled by the software we will write. If we feel the instrument is not responding the way we want it to, we can modify it with minimal effort.

Fifth: Specialized instruments can often be made for a lot less than you may think!

Sixth: *You will develop critical skills you can use the rest of your life.* This in itself might be more important than anything else. (After all, we live in the information age, no?)

We can say with some confidence that with a little learning and effort on our part, we are can create instruments tailored to our needs that will help us be more productive. (And in the process make our lives more interesting.)

So what kinds of things can be sensed easily? We need to know what we can sense because what we sense will be the data source we feed into our microcontroller-based instrument/controller.

Types of Sensors

Inexpensive sensors are readily available for most of the following:

- Humidity
- Gravity
  - Level
  - Acceleration
- Resistance
- Capacitance
- Voltage
- Frequency
- Magnetic field
  - Hall effect sensor
- Pressure
  - Altimeter
  - Vacuum
- Distance
- Sonic
- Thermal
  - Infrared
  - Temperature
- Light
- Chemical sensors
  - PH
- Sound
  - Noise
  - Sound pressure
- Contour or roughness
- Speed
  - RPM
- Position
  - GPS
- Orientation
  - Gravity sensors for two and three axes

We will, of course, use only a couple of these sensors because our interest is not primarily in the sensors but in manipulating the information we get *from* the sensors, no matter what type of sensors they are. In order to do that, we must learn how to effectively connect to whatever is provided to us and provide outputs that can control the devices we are interested in manipulating. One chapter in this book is devoted to making the connections on the input side, while one concerns making the connections on the output side. In the eight projects we will undertake, the following sensor interfacing problems will be addressed:

1. Tachometer	Counting the frequency of a pulse train
2. Metronome	Creating accurate timed intervals (frequencies)
3. Marbles counter	Exploring various counting techniques and how to implement them
4. Dual thermometer	Reading analog signals and displaying them
5. Artificial Horizon	Providing a stable horizontal surface with an unusual instrument
6. Touch screen	Making control panels through a useful technique
7. Single-point controller	Controlling one variable (derived from item 4 of this list)
8. Solar collector	Exploring data collection over a long time period; data logging

These eight projects are designed to familiarize you with the basic techniques needed to build instruments and controllers in today's engineering laboratories and on hobbyist workbenches.

Two Interesting Resources You Will Want to Investigate

RadioShack has a sensors learning laboratory (~$50) that provides an inexpensive way to access a family of sensors that can be used with our instruments.

The McGraw-Hill book *Electronic Sensors for the Evil Genius* discusses a number of interesting sensors and explains how to use them. These techniques, and others that are modifications of them, can be used in the instruments and controllers we shall create.

11

CONDITIONING THE INPUT SIGNAL

General

First, we need to get a handle on the signals coming into our instruments. This chapter covers the techniques used to collect and condition the input signals so they can be connected to the microcontroller.

Before we can do useful work, we need to detect and manage the signal we are interested in. In our particular case, there is the added requirement that the signal either be an electrical signal or be such that it can easily be converted into an electrical signal. Though a Bourdon tube pressure gauge might read the pressure just fine for most engineering applications, we need a pressure sensor that provides us with an electrical signal we can feed into an electrical instrument. We are not so much concerned with the magnitude of the signal as we are with knowing it is in a range we can read, manipulate, and bring into our microcontrollers. A linear signal response is most desirable, but we can take care of the nonlinearity of a signal in our controller with software if that becomes necessary.

In the engineering laboratory and amateur engineer's workshop, the electrical signals we encounter will most likely be somewhere between 0 and 120 volts ac and between 0 and 24 volts dc. No matter what the conditions, extreme caution and care should be exercised when dealing with any kind of electrical signal. Electricity can be lethal if not handled with care. When you are dealing with the higher currents and voltages, seek the help of an experienced electrician if you feel the least bit uncomfortable with undertaking that part of the task. Be informed that certain work must meet the requirements of the National Electrical Code in most jurisdictions, so help from an experienced person (electrician) can save you a lot of time and headaches.

Let's take a look at the ac signals first, and then we will look at dc signals.

Alternating Current Outline

- **High voltage**   120 volts (we will not consider anything over 120 volts)
- **Low voltage**   24 volts
- **Electronic (computer circuitry signals)**   Less than 24 volts, and usually less than or equal to 5 volts

ALTERNATING CURRENT

The signals we are interested in as they relate to the electric grid and laboratory devices will most likely either be at 24 volts ac or at 120 volts ac. Chances are good that if we got it from anything connected to the national electrical grid, it will be in this voltage range. Everything under a half horsepower that can be plugged into an electrical socket is likely to be at 120 volts ac at 60 Hz. The control wiring most commonly used in the United States is at 24 volts ac (but sometimes higher voltages are used). Certain starters, relays, and coils will use 24 volts and will need a transformer to create this voltage. Look for this transformer and actually measure the voltages at the devices. Control wiring does not usually have to be in conduit if it is at 24 volts or below (check the National Electrical Code). If we want to interact with a control signal, we must extend these signals from the device to our controller, condition it, and feed it to the microcontroller. (Output to things needing 24 volts dc or 120 volts ac is discussed later in the Chapter 12.)

The other ac electrical property we are interested in is the frequency of the signal. A number of instruments we are interested in are based on how often a signal changes over a period of time, either at a fixed rate or a variable rate. Variable rates are usually read with counters over an extended period of time, while fixed rates can be read over a very short period. Oftentimes, the accurate measurement of one cycle will suffice. The measurement of pulse widths is a mixed bag in that we measure the time between changes of a fixed signal.

Computer Input   If we are getting information from a computer, we will need an interface that matches the output of the computer both at the voltage level and as a communication protocol. These signals are almost always at or under 12 volts.

Direct Current Outline

- 24 volts (we will not cover anything over 24 volts dc)
- 12 volts
- 5 volts
- 3.3 volts

DIRECT CURRENT

The signals we are interested in are under 24 volts, but before you do anything else you should check this for yourself. If lead acid batteries are involved, it will probably be at 12 volts. Most automotive voltages are currently at 12 volts (the measured voltage on a charged battery on a running car will be slightly higher). Heavy duty trucks often use twice this voltage. If you are considering electronic devices, the voltages will probably be between 3 and 9 volts. Conditioning dc signals for introduction into a microprocessor is usually a matter of making sure not too much current or too high a voltage goes into the electronics with appropriate resistances and buffers.

Note *Safety is always "JOB 1." Be cautioned that when working around lead acid batteries, extremely large currents (well over 500 amps instantaneously) are easily available at the battery terminals. These high currents can be dangerous and will easily melt wires and even wrenches. Be very careful when working around lead acid batteries. If you are not familiar with hazards associated with batteries (acids and hydrogen), you should take the time to become acquainted with this aspect of your work for your own safety, and the safety of those around you, before you start.*

All the signals we collect must be conditioned and converted to a form that the microprocessor can safely accept. The conversion into digital format can take place in the microprocessor itself, so we do not have to worry about any external electronics to do that.

We have two interests in the incoming signal. With dc signals, we want to know if the signal is on or off, and we want to know its magnitude. (dc signals can also come in as a square wave or a frequency, which is covered under ac.)

On-off information is used as a means of detecting what the conditions downstream from us are so we can send control information to what is upstream from us. If a device that is interacting with us has come on, we need to respond to it in a specific manner, meaning that the information should be used as part of our control functions.

Magnitude information is used to determine the condition of the signal we are monitoring, to know if we are getting the results we want. If we are trying to maintain the temperature of an oven, we need to know the magnitude of the signal coming to us across the thermocouple so we can convert it into a temperature and then compare that temperature to what was needed, in order to decide whether to turn the heating elements either on or off. The instrument we design manipulates the thermocouple voltage data, to comprehend the information in the signal, and then makes the decision that controls the heaters.

Simple Switches and Other Contacts

KEYBOARDS

The reading of keyboards and other matrices of switches is covered in great detail in the earlier part of the tutorial under keyboards.

RELAYS

A relay can easily be used to provide a dry contact that can be read with the PIC. These contacts are quite noisy (electrically) and should be debounced either with hardware or software as they are read. If two pins are available for the signal, a latch can be set at one pin and cleared with the other. Latches have the advantage that a latched signal latches immediately and then does not need any more attention. We can come back to it when convenient. A signal latching and clearing circuit is shown in Figure 11.1.

RESISTANCES

Potentiometers can be read by placing them across a reference voltage and ground and using one of the analog inputs to read the position of the wiper. This technique is demonstrated in the chapter on reading inputs and in the final project where a potentiometer is used as the set point determining device.

Fixed resistances can be read either by measuring the time it takes to discharge a known value capacitor or by comparing them to a known resistor. The technique used will depend on the value of the resistor being read. The first technique is explained in the PBP manual under the POT command on page 116. The second technique places the two resistances in series and then reads the middle voltage as it represents the wiper in a potentiometer. The caution is that low resistances are likely to short across the power supply and ground. The total resistance of the two resistors needs to be well over 2k ohms and is better if over 5k ohms.

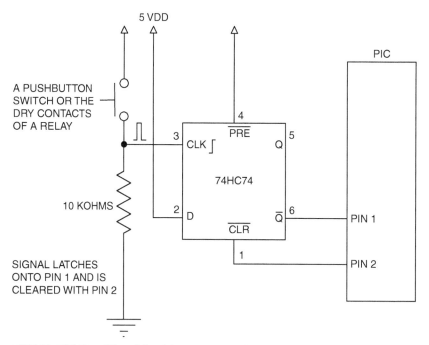

Figure 11.1 Signal latching and clearing. (Reading a relay or a pushbutton with a latch.)

Circuitry for Conditioning dc Signals

24-VOLT SIGNALS

Signals between 12 and 24 volts dc can be introduced into a PIC by creating a voltage divider to reduce the signal to approximately 4.5 to 4.9 volts dc as shown in Figure 11.2. A current-limiting resistor of between 220 and 1000 ohms is placed in series with the signal as a safety precaution to limit the current into the PIC. The PIC inputs are high-impedance inputs but it does not hurt to add the resistance in an experimental situation where things might turn out to be not quite what you expected.

12-VOLT SIGNALS

Signals at 12 volts can be handled the same as the 24-volt signals above with appropriate resistances provided, or they can be passed through standard logic components to condition for use with a PIC. All signals should go though conditioning buffers or gates when there is any concern about the quality of the signal.

The PIC pins are in a high-impedance condition when programmed as inputs and will accept any TTL- or CMOS-level signal, so the task at hand is to convert the incoming signal into either a TTL-level signal or a CMOS-level signal. Floating signals coming

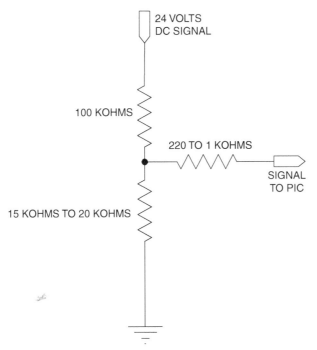

Figure 11.2 Wiring diagram for reading a 24-volt signal.

into floating inputs should have a pull-up resistor (10 to 100k ohms) at the pin to tie the pin high. Opto isolators can be used to isolate the noisy signal electrically from the PIC when necessary. If you have an isolator available, use it.

The diagrams in Figures 11.3 through 11.6 show various techniques that may be used.

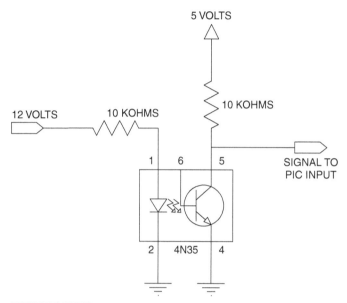

Figure 11.3 Opto isolation of incoming 12-volt signal.

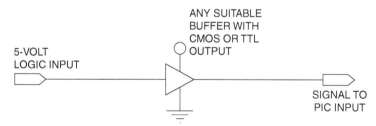

Figure 11.4 5-volt TTL-level signals.

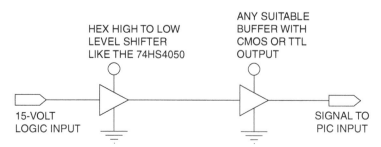

Figure 11.5 15-volt logic.

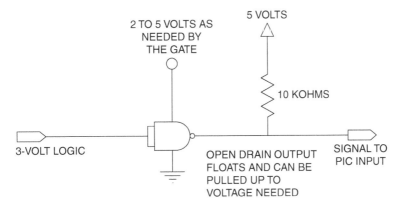

Figure 11.6 3-volt signals.

12

CONDITIONING THE OUTPUT SIGNAL

General

The microcontroller we use will either be a TTL- or CMOS-level input/output device. Its signals can be introduced to other similar devices, almost always, without altering them. However, the real world is not full of TTL and CMOS devices, and we will have to connect to motors and relays, Silicon Controlled Rectifier (SCRs) and TRIode for Alternating Current (TRIACS), lights and pumps, and a host of other equipment on our bench tops and in our laboratories. This chapter shows us how to take the TTL signal as it comes out of the microcontroller and prepare it to inform and control the real world.

The types of devices we will be sending information to can be broken down into the following:

- Computers
  - Serial interfaces
    - COM1
    - COM2
  - Parallel ports are more complicated to use and will not be discussed in detail. (Today, they are not often used for interaction with MCUs.)
- Inductive loads
  - Motors
  - Relays
  - Solenoids
- Resistive loads
  - Light bulbs
  - Heaters

COMPUTER SERIAL INTERFACE

The PIC family of microcontrollers can send information to, and receive information from, a computer. This can be done serially and in the parallel mode. It is much easier to use serial communications, which is covered in detail in Chapter 9.

COMPUTER PARALLEL INTERFACE

Some PICs provide the ability to create a slave parallel port. On the PIC 16F877A, PORTD is used for this, while PORTE is used to control and monitor data transfer.

The datasheets contain details on using slave ports for these interfaces, but they are not necessary at the level of this text, and if all you want to do is communicate with a PC, by far the easiest way to do so is with a serial RS232 interface. We will not cover parallel port communications in this book.

INDUCTIVE LOADS

Any time we are sending an electrical signal to an inductive dc load, we have to make arrangements to absorb the flow of energy released when the inductive load is disconnected. The easiest way to do this is to provide an appropriate diode that will offer a safe path for the reverse current to dissipate. The diode is installed so it is conducting in the direction opposite to the normal flow of current. Attach the cathode of the diode to the positive terminal, and the anode to the negative terminal of the inductive load as close to the load as possible. In this position, no current will flow through the diode when connected across the dc voltage, but when the device (relay, solenoid, and so on) is disconnected, the reverse current from it will flow across this diode and be dissipated in the device itself. Diodes must be selected to match the expected load amperage and voltage. There is no harm in using a diode that is larger than needed.

RESISTIVE LOADS

Resistive loads can be connected directly through solid state relays (SSRs) that will accept TTL-level signals and that can control anywhere from 10 to 40 amps at 120 volts ac directly. These SSRs are the interfaces of choice for connecting these loads because they are easy to use and all the internal electronic work is already done for us (see Figure 12.1). Some of these units even have a tiny LED built into them to indicate when they are active. (Devices often turn off and on when the ac voltage goes through the 0 voltage transition, so you must make sure the load voltage will actually go through 0 volt, or the device will/may not switch.)

Select a unit specifically suited to the task you have in mind. Inexpensive units are available from most supply houses selling surplus electronic equipment on the Internet. I myself have purchased used units and have never received one that did not work.

GENERAL   175

Figure 12.1   A solid state relay. (These solid state relays provide the easiest way to connect to higher voltages. Note LED.)

If you are going to run a motor or other inductive load, make sure the unit has suitable suppressors built into it or provide for the suppression of the transients yourself. You can also isolate the PIC from the load by using a solid state relay to control an appropriately sized mechanical relay, which in turn controls the motor. This scheme would not be suitable for loads that were to be cycled more than one or two times a minute. You will burn the relay contacts out if they are not overly oversized.

13

AN INTRODUCTION TO THE EIGHT PROJECTS

The Web Site

Use the support Web site. It is an important part of all the information you have access to with this tutorial. Specific to the following chapters, it has extensive pictures on it that show how all the projects were built by me, and these can be a tremendous help to you in building your own projects, even if you decide to modify and extend the projects beyond the minimal and basic construction I have undertaken. The pictures are in color and color adds a lot of useful information.

The Eight Techniques

You must become comfortable with eight basic techniques to be able to design and build instruments and controllers that can monitor or control the variables and properties you have an interest in. They are

1. Counting synchronous and asynchronous pulses.
2. Creating accurate short- and long-timed intervals with timers.
3. Using counters effectively.
4. Sensing and reading analog voltages.
5. Using pulses to control external devices.
6. Creating simple scanning routines to monitor phenomena.
7. Controlling the property or function you are interested in.
8. Logging data over an extended time period automatically.

The eight basic constructions in this tutorial cover these techniques one at a time. The projects are designed to give you an expansive view of the many possibilities that using

PIC microprocessors offers you, the experimenter. They selectively demonstrate what can be done by the amateur experimenter and engineering student with fairly minimal resources. This is an introductory tutorial, as opposed to a highly technical treatment you might find in a more rigorous text. The projects are fairly straightforward and are designed to expand your horizons and give you the confidence you need to create and build the instruments necessary for the work you want to undertake. They demonstrate a varied approach to a seemingly random set of problems, which when seen as a whole, give us the experiences we need to move to the next level, which is of course the design and construction of the instruments required.

The eight instruments are:

Instrument Name	**Function Being Studied**
1. Tachometer	Counting pulses (synchronous)
2. Metronome	Timer techniques
3. Marble counter	Counting techniques (asynchronous)
4. Dual temperature sensor	Analog-to-digital conversion considerations
5. Artificial horizon table	Converting pulses to motion
6. Touch screen	A useful real-world scanning application
7. Single-point controller	Controlling a "set point" process with details
8. Solar collector	Data logging over time

These eight projects represent the eight fundamental techniques you must master to be able to build the instruments and controllers you need. Each instrument in the series is designed to isolate and address one part of the data collection or conversion problem. Once we understand the basic components in these systems, we can proceed with the techniques necessary to design the PIC-based instruments necessary for the tasks at hand.

As always, the first thing we must do is to convert the signal we are interested in into a useable digital format. Often, you may need to amplify the signal before converting it to the digital format required.

1. The tachometer project is a basic exercise in understanding the counting of pulses that can come in at widely varying rates. (It also teaches you how to use seven segment displays.)
2. The metronomes have to do with learning how to use the timers to create accurately timed intervals. The metronomes we create will operate identically, but will use the various timers to create the intervals needed. This is a detailed exercise in the use of timers in the PIC 16F877A. Both the LAB-X1 and the tachometer created earlier will/can be used to create the metronomes.
3. In the marble counters section, we will learn about good, bad, and ambiguous signals, and experiment with some of the techniques used for sorting things out with a microprocessor as we collect the data. These counting techniques can and will be applied to all sorts of instruments that you create in the future.
4. The two thermometer instrument allows you to measure two quantities simultaneously. Both are temperatures, but they do not have to be. One can also be designed to be a set point. This is the basic instrument/controller, and the instruments you create will essentially be variations of this project, with the appropriate signal conversion

modules added. Later on in Project 13.7, we will convert this instrument into a controller, and then in Project 13.8 we will use it to log the data from a solar collector over an extended period. It is important to understand that the two quantities considered do not have to be the same thing. We could use the linear brightness of a light to control the parabolic speed of a motor by putting the information through a custom-designed controller.

5. In the artificial horizon project, we learn how to use pulsed signals read from a sensor to control the position of a table positioned with two model aircraft servos. The servos are fed signals that are a function of the error signal read from the sensor and hold the table in a horizontal position.
6. Building the touch screen is the basis for learning about scanning routines. Though this is a relatively small surface, we can demonstrate and learn about the techniques used to make a touch screen with this project. A touch screen is a control panel you can make on the laboratory bench with minimal cost and effort.
7. The two-input controller is a finished instrument that incorporates the competencies learned in the previously mentioned projects. We then use this instrument in the next project to create a data logger.
8. Data logging is covered in the context of a solar collector and uses the two-input controller to provide the data that we then log over an extended period of time automatically. The data will be sent to a personal computer for storage and go through eventual analysis at a later date.

Adding a little more detail to the preceding descriptions, we will discuss, then build, and hopefully understand the engineering and science behind the following eight projects.

1. THE BASICS OF COUNTING PULSES: THE TACHOMETER PROJECT

Research has indicated that carefully managing the engine speed as you drive around town can substantially increase the efficiency of your automobile. Design and build an inexpensive tachometer that can be added to an automotive engine with minimal effort. Design the device to display the engine rpm (revolutions per minute) on a four-digit display. The display and the CPU board are to be placed at some convenient location in front of the driver. The system must use the 12 dc power available on the automobile and start when the ignition switch is turned on.

This is an exercise in counting pulses that come in at various rates, and displaying the results on seven-segment displays or on an LCD. Almost all the signals we use with microprocessors end up having to do with counting and manipulating pulses, so this is a core competency you must master.

2. THE METRONOME: CREATING CONTROLLED PULSES

Ms. Music, our local high school music teacher, did not get the funding for the metronomes she wanted for her class. The principal has asked your electronics instructor so see if he/she can get the class to create 25 low-cost metronomes for the music students.

The electronic metronome is an exercise in creating accurate time intervals that are controlled from a potentiometer on both the LAB-X1 board and the tachometer board. The exercise includes programming for both the LCD on the LAB-X1 and for a display consisting of the 4 seven-segment LEDs used on the tachometer in the first exercise.

3. COUNTING MARBLES: DISCRIMINATING BETWEEN VARYING PULSED SIGNALS AND FEEDING THE TIMER/COUNTERS

Part 1... Mr. Marbles the manager of the local marble factory has been plagued with marble counting problems and has approached the head of the electronics department to see if he can come up with an inexpensive solution. The class is charged with designing a marble counter that counts the marbles as they go by single file at the rate of about ten marbles per second, and the factory manager wants to increase that rate as much as possible. The counter that can count the highest number of marbles in 10 seconds accurately will be selected as the instrument of choice by the plant manager.

This is about learning how to count the pulses we encounter when processing information. Some come in slow, some come in fast, some come in very fast, and some are hard to discriminate from the background noise. This exercise exposes you to these real-world problems.

Part 2... Mr. Marbles, our friend from the marble factory, is so happy with the performance of the counter we made for him earlier that he has come back to us for an even faster counter. "What can you make in the way of a really fast counter?" he asks. We have convinced him that we can make him what he would consider a really fast counter, but the marbles need to go by single file just as they did for the previous counters. To this he has agreed.

This instrument will count how many common everyday glass marbles go by its so-called gate. We are doing this just to do it! Someone might just have a use for this, or a modified version of this, but our purpose here is to understand the detection, amplification, and the counting of small seemingly random signals coming in helter skelter.

In this experiment, things start to get complicated and we have to use our wits to figure out how we can solve the problems. So, in a way, this is an exposure to the real world.

4. THE DUAL THERMOMETER: TWO ANALOG SIGNALS CONVERTED TO DIGITAL AND DISPLAYED

Dr. Thermo is in the process of undertaking a large research project to check the energy transfer across a large number of coils to be used in the air-conditioning industry. He has indicated that he needs to know how the temperature of the air and the temperature of the cooling brine changes as the heat exchange takes place. Design an instrument to display the inlet temperature and the outlet temperature for both fluids.

The instrument is an electronic thermometer that reads two temperatures simultaneously. This is an instrument you can use every day in your day-to-day monitoring of the systems around you. Each sensor costs about $3. This exercise is about reading analog signals, converting them to digital format, and then interpreting them for display on a two-line liquid crystal display. With this instrument you can determine the energy flow in most systems if you know the rate of flow and the mass properties of what is going by.

This project contains the basic techniques you will need to master to process analog signals as opposed to digital signals.

5. AN ELECTRONIC ARTIFICIAL HORIZON: SOPHISTICATED INSTRUMENTS MADE EASY WITH MICROCONTROLLERS

Parallax sells a very interesting gravity sensor that indicates the change in gravity in both the X and Y directions as well as the ambient temperature at the instrument as three frequencies. (We will not use the temperature sensor part in our experiments.) The sensor can sense up to 2 Gs of acceleration with a surprisingly good resolution of 0.001 G (a milli-G). We will use this device to create a two-axis table that stays horizontal while the surface it is mounted on is turned every which way (within about 20 degrees) relative to the horizon.

The accelerometer we will be using is the Memsic 2125. The Parallax company delivers it already mounted on a tiny board with pins at 0.1 inches on center. It plugs directly into an experimental breadboard with standard 0.1 inch on-center holes. There are only six pins, and two of them are at ground! Three other lines provide the three frequencies we are interested in, while the sixth pin is the power pin. How simple can this get?!

The signals from this sensor are in the form of frequencies that vary with the tilt of the sensor in the X and Y direction (and the ambient temperature). The sensor is most sensitive near horizontal, and least sensitive when the sensor axis is held vertical. We can take the frequencies received and process them so they give us the signals we need to control a couple of hobby R/C servos. The linkages between the servo and table are to be arranged to keep the table approximately level. The final correction can be provided in software with a lookup table but that may not be necessary depending on how we design the software!

6. THE TOUCH SCREEN

The touch screen teaches us how to scan a number of signals to decide what needs to be done under circumstances controlled by the signals. Useful touch panels can be created to simulate the operation of control panels for electronic devices with minimal expense. In this exercise, we create a touch panel that controls the blinking rate of two LEDs and displays the conditions at the panel on the LCD.

Simple touch panels can be placed in front of simple graphics to create the inputs we need to control the instrument and controller we are building.

7. THE FINISHED CONTROLLER

In this project, we convert the two-temperature thermometer into an adjustable thermostat with an external inhibit capability to demonstrate our ability to create a controller/instrument.

8. LOGGING DATA FROM A SOLAR COLLECTOR

Now that we have a working two-thermometer sensor, we add communications to it and create a system that logs the conditions inside a solar collector, to a PC, automatically every minute (or another time interval) for a year.

A large portion of the tutorial is devoted to each one of the preceding devices, and building instructions as well as a discussion of the theory and the programming techniques used are included in each chapter.

Notes

In the following exercises, we will rely as much as we can, on the capabilities of the PIC and try to do the projects with as few external components as possible to keep our costs down and to increase our knowledge of using the PIC to do whatever it is capable of doing. For example, in the first project, the tachometer, we could easily have used ICs that convert BCD (binary coded decimal) data to what is needed by the seven-segment displays to display a number, but we did it without the ICs to learn how this can be done with the PIC alone.

On the software end, we will use as few of the instructions from PROBASIC as possible to keep the emphasis on the design of the instruments as opposed to mastering the tricks and techniques possible with the extensive software language.

The emphasis is on the PIC and what is inside it.

STOP! STOP! STOP! STOP! STOP!
ONCE IN A LIFETIME CHANCE
TO EXERCISE YOUR WITS!

You now have a rare opportunity to exercise your creative thinking!

Now that you know what we will be building, you can substantially increase what you will get from these exercises by not reading the following pages for a couple of days and spending time thinking seriously about how you would use the resources available to you to design the instruments described in this chapter.

Make drawings and sketches of what you would do in your shop log. Later, it will be instructive to compare what you envisioned with how I have described the instruments. Hopefully, what you come up with (in at least some cases) will be more inspired than the simplified methods and techniques offered in this tutorial.

Good luck.

14

THE UNIVERSAL INSTRUMENT: A BACKGROUND DISCUSSION

The Properties and Capabilities of a Universal Instrument

Almost all the instruments we will make have some shared characteristics. What are they and what advantages and disadvantages do we get from them? What does having our instruments based on PIC microcontrollers do for us? Asking these questions and understanding basic concepts like these makes it easier for us to proceed with our work.

Let's look at the properties needed in an instrument by comparing it with an instrument we are all familiar with: a volt/ohm meter or VOM. We will compare its properties and functions to the instruments we are designing to get a better understanding of what we are trying to accomplish and how we might accomplish it.

Each instrument/controller must have the following basic properties:

1. It must have a way to *enter information* into the device in a convenient manner. In a VOM, we are provided with two probes that we place at various points in the circuitry. In the instruments we make, we will provide connection points for the signals and pulses we are interested in reading, but more importantly we are interested in providing inputs that can be *programmed* to influence the operation of the instrument *in real time*. With a little imagination, the instruments we create can be both interactive and intelligent.

2. It must have some form of *decision-making* capability that allows it to manipulate the information sent to it, and some of this capability needs to be controllable (influence-able) by the operator. In the VOM, there is a selector switch that we use to choose ranges and functions we will use. In our instruments, we will use the computer to set the parameters within which our instrument will operate. We will be able to program the instrument for the task at hand as needed. For our instruments, the computer is the selector switch! (Note that in the newer VOMs, some autoranging capabilities are implemented for selecting the range used by the meter. The output, too, is ranged so it can fit in the limited number of characters in the display. To us, this means some of the decision making we might have undertaken has been relegated to the intelligence built into the machine.) The flexibility of *our instruments* is rooted in the ability to reprogram the instrument with a few keystrokes whenever needed. In order to be able to use this flexibility, we must design the hardware for our instruments in a way that will lend itself to software modification with relatively minimal effort in a relatively short time. We have to think about what the instrument/controller will/can do as we design it so it will perform its intended functions with some elegance. Always keep in mind that the instrument will do only what we design it to do. *You must commit this concept to your brain.* It is *very* important.

3. The device must have a way to *output information* in a useful format. The format can be anything from a simple switch closure to a comprehensive CRT display. On the VOM, a few digits suffice to give us the information we need. On instruments based on PIC microcontrollers, the addition of a two-line display can provide a considerable amount of information. The many output port lines can be used to turn LEDs on and off and to provide other on/off signals. A serial port can be used to send and receive information from a computer. Audio signals can also be added. Versatility will be our strong suit.

4. The information/output desired should be able to be *stabilized* for a while so we can collect it. Some of the newer VOMs have a storage function that lets you freeze the display at selected peak values or at selected times. Our instruments will be able to do that and send the information to a computer at selected intervals for storage in that computer to facilitate future analysis. They will be able to *make decisions in real time* when we are not there, and take corrective actions, or summon human intervention. The capabilities we can incorporate into our instruments are limited only by our imagination. We are the designers.

A Basic Temperature-Controlling Device

Next, we will discuss this in the context of a simple temperature-controlling device. Later, we will actually build this device.

What are the basic properties and components of an instrument that can be described as a universal all-purpose single-point controller? Basically, the following functions are called for:

- Sense condition
- Set a point of control
- Compare the preceding two functions
- Output a signal

Let's go over them one at a time.

1. A simple universal controller would allow us to control one set point or property. For the purpose of this discussion, we are assuming this is a temperature. The instrument must allow us to control a temperature. In order to do this, it must have a connection for the temperature sensor, and what we provide will depend on the sensor we have selected. We will use an LM34 temperature-sensing IC as our detector. This device uses a three-wire connection, so we will provide a three-screw terminal for this connection.
2. The device must have an adjustable setting variable of some sort to which the above temperature is to be compared. This is the output set point control device. For our instrument, we will use a potentiometer. Once we have the two signals in our instrument working, we can make a decision as to whether we want to turn on the heat or the refrigeration. This will be determined with the software we will write to control the instrument. (The potentiometer is a three-wire voltage divider that we will use to parallel the input that the temperature detector is providing. This is just like what we discussed earlier in this book, and will make the comparisons easier.)
3. We need a way to turn the actual instrument function on and off. This switch is not the on/off switch for the instrument but another switch that allows us to activate the operational portion of the instrument when we are ready to start the instrument. This switch lets us play with our setup and get everything just right before we activate the instrument. A simple SPST switch will be adequate for this and we will use a small toggle switch for it. Since we may want to mount this switch at a remote location for some of our applications, we will connect the switch to the board via two screw terminals. The major function of a switch like this is to allow our instrument to be controlled by another electronic device! This is a very powerful capability and is the first step in the extended automation process.
4. The controller needs a couple of connectors that the output signal appears on. We will connect the controlled machine component (heater, relay, and so on) to this point. If we have to amplify the control signal, this is where the relay (transistor) or whatever will be connected. If we later want to reverse the signal from normally off to normally on, this is where we program the logic to do so. This is the output of the instrument. We will annunciate the signal with an LED and use two screw terminals as the connection points.
5. The instrument needs some sort of indicator on it to tell us when the output from the instrument is active. This is a convenience item that makes the instrument/controller more user-friendly. We will add an LED to indicate this on our controller. When this LED is on, the instrument is putting out a signal at the output connectors. This LED also lets us see the instrument operating without having to connect up to a real control point, and so on.

The preceding describes almost all instrument/controllers. No matter what you are asked to control, the instrument you design will have the preceding basic modules or components in some form or another. It may control one function, or it may control a hundred functions, but each function will support the previously mentioned abilities.

The controller mentioned earlier is an extension of the thermometer we will construct in one of the projects. Its construction is covered in detail.

Notes

All instruments are assembled from standard hardware and software components, so they tend to have a modular structure. Each module in the instrument provides one function. The following are some common modules you would find on systems based on the PIC 16F877A:

- Analog input section
- Mathematical manipulations
- Set the DEFINEs for the LCD
- Display to the LCD
- Read a potentiometer or two
- Display to displays (CRT)
- Read a keyboard
- Put out a tone
- Read a switch
- A communications module

We create our instruments by selecting or designing the modules we need and assembling them into a project. Once you have created a module or subroutine, you can use it again and again in your designs if you create it to be easily transportable to other designs. Taking the time to do so now will save you a lot of time, again and again, in the future.

Notice in the projects that follow that I too use the modules over and over again. I also made a concerted effort to use as few instructions in PIC Basic Pro (PBP) as possible, so the emphasis was on the design of the instruments, not on fancy software tricks and routines.

15

COUNTING PULSES:
A PROGRAMMABLE TACHOMETER

Project 1

In this first project, we build a programmable tachometer that can detect between 1 and 9999 pulses a second (or another time period) and display the results on 4 seven-segment LED numeric displays. These pulses can be counted in two ways. If we know we are getting a fixed duty cycle square wave, we can measure the total length of a pulse and determine the frequency of the signal from that. If the pulses are not coming at a regular rate, we can actually count the pulses over time and determine the rate that way. We can write the software for both techniques so we can see the difference. At the end of the project, we will take the source of the pulses into consideration and discuss how to modify the software to display an actual rpm (revolutions per minute).

When working with microprocessors, much of the information we deal with is read and processed as pulses, counts, frequencies, and the like because microprocessors are digital devices, and are usually not designed for handling analog values. So it becomes important that you be comfortable with gathering, processing, and displaying information of a digital nature, and this project focuses on that aspect of our knowledge and understanding regarding microprocessors. Once you are comfortable with the design and construction of this tachometer, you will be able to handle all other frequency input projects with confidence.

We will use the LAB-X1 board as a helper instrument as we go along. First, we will make the software work on the LAB-X1 (when we can) and once we know we can read, count, and display the inputs the way we want, we will move on to the fabrication of the seven-segment LED-based tachometer shown in Figure 15.1.

In this project, the LAB-X1 will be particularly helpful because seven-segment displays require a lot or wiring (that can go wrong) and it can be difficult to get things sorted

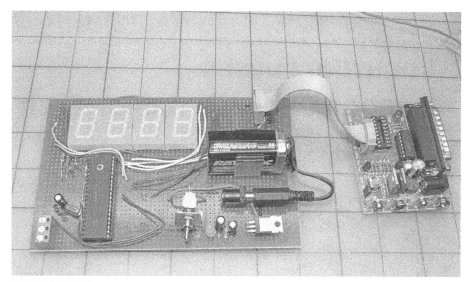

Figure 15.1 A programmable tachometer.

out at times. When using a LAB-X1, we know we have a working platform to develop the software without having to worry about hardware problems that we might have created for ourselves. After we have the tachometer software working, we can use the LAB-X1 to create pulses generated at known frequencies to check the accuracy of the instrument we created.

The displaying of the counts on 4 seven-segment displays has to do with learning how to use these commonly used and inexpensive displays. It is, of course, much easier to display the information on the 2 × 20 LCD on the LAB-X1, but in this exercise we will be using the LAB-X1 only as our initial software platform. If you prefer, the project can easily be modified to run on the LAB-X1, but then you will miss out on the experience of learning how to use the seven-segment displays. In any case, the PIC we use will be the 16F877A. (This PIC will be employed for all the projects in this book, and the LAB-X1 will be used to support all the projects as needed.)

The fact that we are using only four displays is because the integer math provided in the compiler can handle only 16-bit math (65535 max), which limits us to four displays if we want to go to a 9 in the left-most display. It also simplifies our problem at this stage and will demonstrate that refreshing four displays takes us to the maximum that the PIC can control (at 4 MHz) without having to use assembly language programming. It is also a compromise that has to do with the fact that four digits can be handled easily with one and a half ports (12 lines).

We will be using a small dc motor with a 20-slots-per-revolution encoder (available from my Website encodergeek.com) on it as our signal generator (see Figure 15.2), but before we do that we need to undertake some general discussion about counting pulses and what the problems related with low and high "counts/time period" are. If you have some other device that can generate pulses, you can use it if you like.

Figure 15.2 A small dc motor with open encoders attached for generating signal pulses. (Fewer encoder slots give you more time to count them. Open construction gives full access to electronics.)

Figure 15.3 shows a circuit used to make a simple and inexpensive pulse generator with a PIC 16F819. This device can be used to generate the many pulses and intervals needed for the numerous devices that will be created. The fact that it is programmable is the key. No program is provided because you now have the skills to program this device for the outputs you need. The following are a couple of hints. The three-pin output can be programmed to emulate an R/C hobby output, while the 5 volts that a servo needs are provided at the middle pin. The five pins connected to PORTB should be pulled up, and then as they are shorted to ground with jumpers, they select the type of output the signal generator will create. Use the three LEDs as indicators to show you what is going on in the program.

Figure 15.4 shows a picture of the finished device. Bare boards and kits for this are available at encodergeek.com.

PULSE RATE: LOW RATE CONSIDERATIONS

Let's think about this in terms of a small encoded motor like the one shown in Figure 15.2.

First, let's look at what the problems at the low rpm end of the spectrum are. Let's assume we are looking at signals coming from a rotating shaft and want to know the speed of the shaft in rpm. The lowest speed of the shaft is 1 *rps* (revolutions per second) and we want to update the rpm display every 0.25 seconds. We need to be able to detect at least five counts every 0.25 seconds to indicate that the shaft is turning. If the shaft had 20 markers on it, we would have to look at the shaft for at least 0.25 seconds to get these five counts. We need a certain amount of time to process and display the information, so the count needs to be read four to six times a second if we are to meet our minimum specification. We have agreed that there are 20 markers on the shaft of

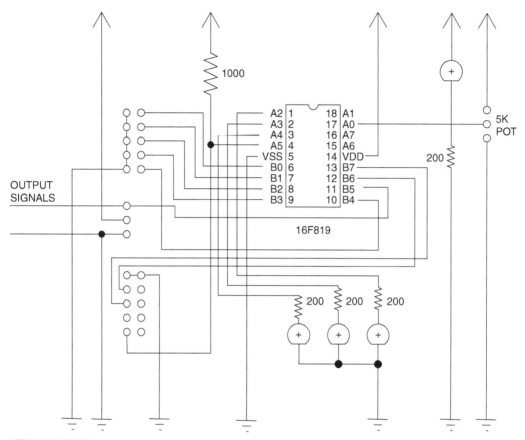

Figure 15.3 Pulse generator circuit diagram.

this motor. We can read these with a photosensitive detector, and the output from the detector is suitable for direct input into our PIC/MCU. This is a reasonable assumption in that most phototransistors will feed directly into a PIC port. (The encoder on our motor meets this specification.)

For low-speed accuracy, the 20 markers must be placed accurately around the shaft. At high rpm, this would not matter because we would be reading many revolutions every 0.25 seconds and the relative placement of the 20 markers around the shaft would be integrated out over time.

We can use the COUNT command in the PBP language to do the counting. For a 0.25-second period, the command would be:

```
COUNT PORTA.2, 250, W1
```

where *PORTA.2* is the line the signal is coming in on
250 is for a 250-millisecond or 0.25-second counting interval
W1 is the variable the count will be placed in

Figure 15.4 Photo of pulse generator.

The PIC will actually take 0.25+ seconds to execute this command (because of the counting time specified) and there is nothing else that can be done while this command is being executed (because of the way the compiler is designed). Since we could get a feel for the rpm in a much shorter time at higher rpm, this can definitely be considered a low-rpm problem. The solution to this low-count situation is to have a higher count encoder on the shaft so a large number of counts can be intercepted in a short period of time. However, as we will see, there is a trade-off at the high speed end when we increase the encoder count.

HIGH SPEED CONSIDERATIONS

There is also a limit to how high the frequency can be and still be in the range that the PIC can respond to. For the PIC 16F877A that we are using in combination with the PBP software, this frequency is 100 kHz for a 20 MHz OSC and *25 kHz for a 4 MHz OSC (our case)*. If we were using an encoder that had 1000 counts per revolution, we could keep up with a shaft that was rotating at 25 revolutions per second.

```
25,000/1000 = 25
```

A motor running at 3600 rpm, which is quite common, spins at 60 revolutions per second. We can see that we start to exceed the limit of what we can do rather rapidly.

So on the one hand, high-count encoders are desirable at low speeds; on the other hand they become a problem at high speed. A corollary to this is that it is very difficult to run a shaft accurately at low speed without some sort of gearing or belt reduction. Think about the problems you would encounter trying to run a shaft at 1 revolution per year accurately directly with a motor.

If our specification calls for a maximum speed of 3600 rpm our encoder *cannot* have more than 300 counts per revolution if we want to maintain the counting time. Using a shorter time will help on the high rpm end but will make things harder in the low rpm region. Of course, we can also use an encoder with many fewer counts and have a perfectly good tachometer.

The software we are using employs integer math, and the largest variable we can use is a 2-byte word. The largest number 2 bytes can accommodate is 65535. This means that in the COUNT instruction:

```
COUNT PORTA.2, 250, W1
```

The number that ends up in W1 cannot exceed 65535. We can accommodate this requirement by shortening the 250 millisecond (or less) time frame, but then we will have problems with the 60-rpm end of the specification. The solution is to use the lowest count that will serve our purposes on the shaft encoder at the lowest speed we are interested in. If we use the 20 counts per revolution we discussed earlier, we will be getting (20 * 3600 / 60) = 1200 counts every second and a quarter of that every 0.25 seconds or 300 counts every second. This indicates that we could use an encoder with a few hundred lines per revolution if accurate slow speed indication was an important consideration. However, adding a higher count encoder to an existing shaft may not be trivial, and adding a few (even 1) equally spaced markers onto the shaft manually with a little epoxy or paint may be adequate for what we need.

There is also the possibility that we could write software that would use different routines for different speeds of the signal, but for this project let's keep it simple. Such sophistications can be added after we get proficient at doing the work at hand. The most important thing to keep in mind is that you must understand the problem in a comprehensive way before you can create a solution.

DETECTION

Next, we must consider the components and circuitry needed to actually react with the signal we are trying to measure (or collect). The simplest way to do this is to react to the changes in the light intensity either as reflected from markers on the shaft or as a disturbance of some other kind in the vicinity. Hall effect sensors are a popular way of detecting rotation in a dirty environment because they are not affected by anything other than magnetic fields. The signal must be converted to a TTL/CMOS-level signal that goes high and low reliably with every change in the stimulus. Figure 15.5 shows one way to create such an instrument interface for the hall effect device.

Since the input to the PIC are Schmidt triggers, we would not normally have to condition the signal for bounce and jitter.

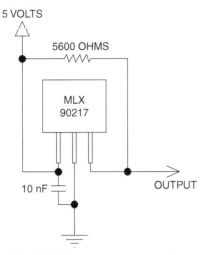

Figure 15.5 From datasheet: Circuits for input from a Hall effect sensor into a PIC. (Melexis MLX90217.)

We will be using the input from a two-channel optical encoder attached to a small dc motor as our signal source. This encoder has 20 slots around its periphery, so we will get 20 pulses per revolution. We will look at just one channel because we are interested only in the signal frequency.

The circuitry for the LED-phototransistor pair for the one channel in this detector is as shown in Figure 15.6.

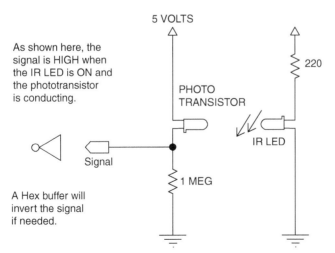

Figure 15.6 Circuits for the input of a tachometer signal into a PIC from the motor optics.

THE SOFTWARE

Let's write the code for the LAB-X1 first and make sure we can read the signals, count them, and display them on the LCD. This lets us build the confidence we need to proceed to the next level. Then, we will build the tachometer as a stand-alone device with the 4 seven-segment displays for output.

First, as usual, let's create the code to enable the LCD. See Program 15.1.

Program 15.1 Counting optical encoder pulses to the LAB-X1

```
CLEAR                       ;
DEFINE OSC 4                ;
DEFINE LCD_DREG PORTD       ; Define LCD connections
DEFINE LCD_DBIT 4           ;
DEFINE LCD_RSREG PORTE      ;
DEFINE LCD_RSBIT 0          ;
DEFINE LCD_EREG PORTE       ;
DEFINE LCD_EBIT 1           ;
LOW PORTE.2                 ;
ADCON1=%00000111            ; this must be set if you are using the LCD
PAUSE 500                   ; it makes the A & E analog ports digital
                            ;
; Next set up the I/O and the variables we will use.
TRISC = %00000011           ;
TRISD =%11110000            ;
PORTD =%00000000            ; Turns the bits off
                            ;
ALPHA VAR WORD              ;
RPM VAR WORD                ;
                            ;
; Then display the LCD message to tell us we are ready
                            ;
LCDOUT $FE, $01, "Tachometer Ready"  ; on the first line.
PAUSE 250                   ;
                            ;
; The loop that displays the count status for the tachometer is
; now ready to be created.
                            ;
LOOP:                       ; the display loop
COUNT PORTC.0, 250, ALPHA   ; read the counts into the PIC
RPM=60 *ALPHA * 4 / 5       ; convert counts to rpm
LCDOUT $FE, $C0, DEC4 RPM,"      "  ; print on line 2
GOTO LOOP                   ; do it forever
END                         ;
```

It's that simple with the LAB-X1 and the PBP compiler.

Next, let's design the software for the 4 seven-segment displays. Four of them will allow for a maximum display of 9999 (but since we are taking samples for 0.25 seconds, and we have 20 counts per revolution, the maximum rpm we can recognize without reducing the sampling time is 119,400).

Max count: 9999 in 0.25 seconds

39,996 in 1 second

There are 20 counts per revolution, which gives:

1,999.0 revolutions per *seconds*, or

(39,996/20/)*60 revolutions per *minute*

We can detect (but not display) a maximum of 119,940 rpm if we count 20 counts per revolution for 0.25 seconds. If we want to detect and display a higher rpm, we must reduce the time interval in the COUNT instruction or decrease the encoder count.

We have already set up PORTD with the lower 4 bits as outputs (but if we had not we would have to add the line of code to do that). After inserting an appropriate decision-making instruction, we can use any of these pins for an output signal to an appropriate device.

If we amplify the signal appropriately, as covered in Chapter 5, the signal on any of the output pins mentioned earlier can be used to control a heater or pump or whatever else we have in mind.

SEVEN-SEGMENT DISPLAYS

Seven-segment displays come in two types: common anode and common cathode. In both types, one side of all the LEDs is wired together to make a common connection. In our experiments, we will use common anode (CA) displays only (see Figure 15.7).

Most so-called seven-segment LED displays actually have eight or nine LEDs in them. The eighth and ninth LEDs are the decimal points on one or two sides of the number displayed. We will not use the decimal points for this project, so we will need only seven lines, but we will wire in the eighth line of the port to activate one decimal point of each display for possible future use.

First, we need to understand a little bit about how seven-segment displays are used. In a seven-segment LED display, all the segments have one common leg, the one that will go to the anode. The data is impressed on the other ends of the LEDs, and when the common leg is connected to the power, whatever data is impressed on the segments gets turned on. The segments are turned on one at a time in most cases, but they can also all be turned on simultaneously. Often, there will be more than one anode pin on a display, but all of them will be connected together inside the display. These extra pins are provided to make it easier to lay out the wiring on a printed circuit board.

All the display segments for each of the displays are connected in parallel to the data lines to one PIC port, and each of the common lines is assigned to a separate line from the PIC on another port. Each of the seven-segment modules is energized one at a time in rapid succession so that it seems to the human eye that they are all on at the same time. Deciding how long each segment must be lit and how often the display must be refreshed can be done on a trial-and-error basis till you come up with the best display possible under the circumstances. The segments are all the same size, so they require the same current, but the decimal point is smaller and needs less current. It may be necessary to power it for a shorter time or provide a larger resistor for it to match the brightness of all the other segments.

196 COUNTING PULSES: A PROGRAMMABLE TACHOMETER

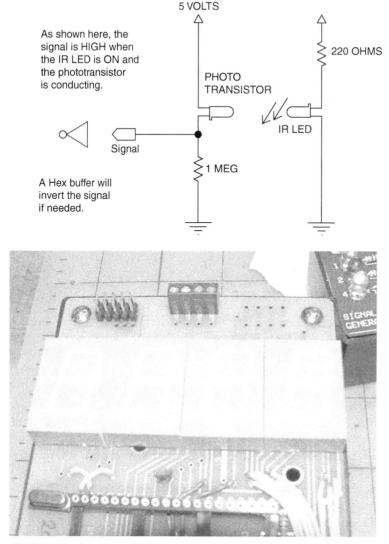

Figure 15.7 A common anode seven-segment display.

One way of laying out the circuitry for a common anode four-character display would be as shown in Figure 15.8. We would use a contiguous port to connect to the eight segment lines and a half port (for the four units in our case) to select the display to be lighted using one line at a time, again in rapid succession.

As mentioned earlier, the common line is the common anode (+). The connections and the program code that control the segments must be adjusted to suit the devices selected. Even though we did wire it, we will ignore the decimal point for this particular application. However, we will test it to make sure it works.

How the data lines are wired (the wiring sequence) is not important because whatever is needed for the display desired can be described in the software and does not make

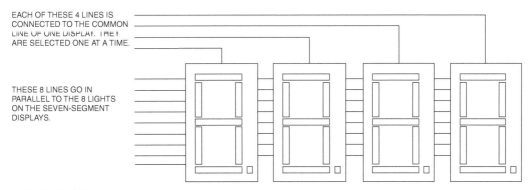

Figure 15.8 Schematic layout: wiring for a set of four seven-segment displays.

any difference to the wiring of the electronics. Once the segments have been wired, the scheme needed to display the various numbers on the segments can be figured out.

In Figure 15.8, the lighting scheme for the displays is as defined in the following meta code:

Execute the COUNT instruction. The COUNT instruction tells us how many pulses were detected. We cannot access the displays while this count is being read. This causes a blinking of the displays on a machine running at 4 MHz. Even when we use an interrupt-driven routine to light the displays, this does not solve this problem because interrupts are not serviced within an instruction in code generated by the PBP compiler.

Convert this to an rpm

Get the **first** number separated out

Impress the first number on the data lines

Activate the first common line as you cycle through the segments

Activate the seven segments one at a time

Get the **second** number separated out

Impress the second number on the data lines

Activate the second common line as you cycle through the segments

Activate the seven segments one at a time

Get the **third** number separated out

Impress the third number on the data lines

Activate the third common line as you cycle through the segments

Activate the seven segments one at a time

Get the **fourth** number separated out

Impress the fourth number on the data lines

Activate the fourth common line as you cycle through the segments

Activate the seven segments one at a time

Go back to the beginning to do it again

There is one big difference between displaying on the LCD and displaying on the seven-segment displays. Once you have written to the LCD, the information on the LCD stays on the LCD. You do not have to do anything else. However, on the seven-segments displays, the information must be updated constantly. This means the driving routine must be based on an arrangement that is called often enough to make the display look like it is on all the time.

The information you need to lay out the wiring to implement the use of the 4 seven segment displays is provided in Figures 15-9 to 15-11. Being able to look at the layout from above and below makes it easier to visual the need of the top and bottom of the PC board layout.

The actual code segment for the wiring shown in Figure 15.8 is laid out in Program 15.2.

Program 15.2   Programming segment for 4 seven-segment displays

```
DISPLAY:                    ; reads each digit and then
  READ DIGIT1, VALUE        ; select first digit
  PORTA=%00000001           ; turn on the first common line
  GOSUB SHOW                ; turn on the segments
  PORTA=%00000010           ; turn on the next common line
  READ DIGIT2, VALUE        ;
  GOSUB SHOW                ; turn on the segment
  PORTA=%00001000           ; turn on the next common line
  READ DIGIT3, VALUE        ;
  GOSUB SHOW                ; turn on the segments
  PORTA=%00100000           ; turn on the next common line
  READ DIGIT4, VALUE        ;
  GOSUB SHOW                ; turn on the segments
 RETURN                     ; end of subroutine
                            ;
SHOW:                       ; shows each segment one at a time
  Z=%11111110               ; selects one segment at a time
  FOR X=1 TO 8              ; do the 8 segments, includes dec. point
    Y=VALUE                 ; get value to show
    Y.2=1                   ; inhibits the decimal point
    PORTB=VALUE | Z         ; makes value and z select one segment
    Z=(Z<<1)+1              ; go to next segment
    PAUSEUS P               ; pause to show segment
  NEXT X                    ; next segment
 PORTA=0                    ; clear PortA
 PORTB=0                    ; clear portb to prevent ghosting
 PAUSEUS P                  ; pause to show clear segments
 RETURN                     ; end of subroutine
```

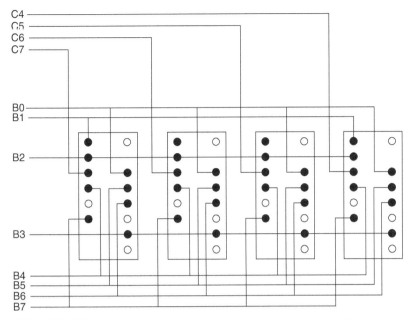

Figure 15.9 One way of wiring for the seven-segment displays as seen from above.

The display routine is written as a self-contained subroutine so it can be called by a relatively frequent interrupt to keep the display alive while a long and complicated program goes about doing its work.

Program 15.2, with a bit more code, replaces the code for the LCD in the original program. The complete program is listed after the wiring diagrams and some seven-segment related notes that follow.

Notes on Using Seven-Segment Displays

The lighting of the displays is somewhat like the reverse of scanning a keyboard. The segments that are lit represent the reverse of the keys that were pressed. In either case, the scan identifies either a displayed set of segments or a key. Most scanning routines serve similar functions and are implemented in this same way.

Let's write a short routine to light all the segments in a seven-segment display one segment at a time. Assume that the seven segments and the decimal point are connected to PORTB. The common line is attached to PORTA.1. Let's assume a common anode

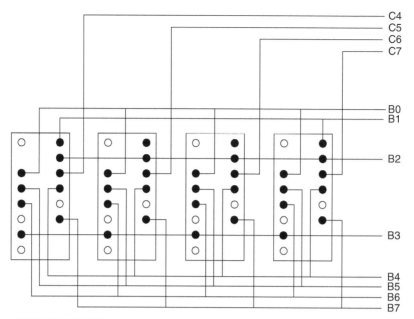

Figure 15.10 One way of wiring for the seven-segment displays as seen from below.

display. A current limiting resistor of 220 to 470 ohms is needed between each segment and the common anode.

```
PORTA=%00000001   ; turn on the first segment
PORTB=%11111110   ; select the first segment
FOR X=1 TO 8      ; there are 8 segments
  PAUSE 100       ; Look at each segment
  PORTB=PORTB<<1  ; go to next segment
PORTB=PORTB + 1   ; put the 1 back in PortB bit 0
NEXT X            ; do the next one
```

In the preceding program clip, we start with the segment connected to B.0 ON and shift the bits one position to the left with each iteration. Adding 1 to PORTB replaces the 0 in the bit after the shift to the left with a 1. The last iteration leaves %11111111 in PORTB and turns all the segments off. If we want to suppress the decimal point, we have to add a line of code to place a 1 in the affected bit after each iteration. Assume that the decimal point is connected to bit 3. Adding:

```
PORTB.3=1;        ;
```

after the shift instruction puts a 1 back at bit 3 and keeps the decimal point turned off.

The finished working program (see Program 15.3) uses the pulse length determining instruction to calculate the pulse rate.

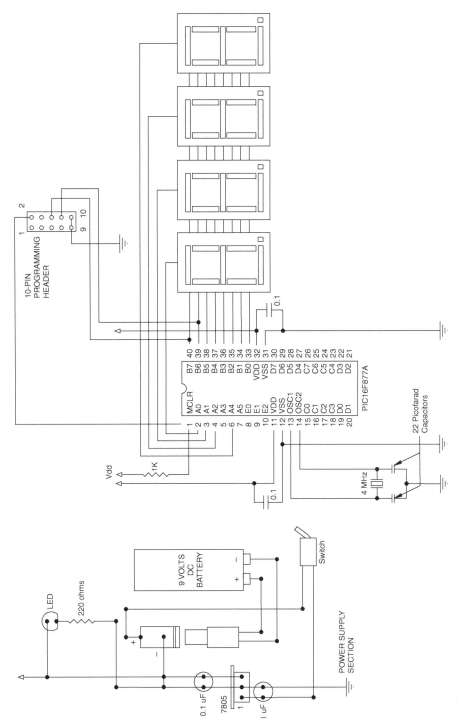

Figure 15.11 Wiring diagram for the programmable pulse counter.

201

Program 15.3 Finished pulse counting program

```
CLEAR                       ;
DEFINE OSC 4                ; Oscillator speed
                            ;
; The following are the images of the numbers and are stored in
; memory
; before we do anything else. These will be placed in PORTB to
; display the
; number for each digit.

WRITE 1999, %11111111       ; blank, no LED is lit
WRITE 2000, %00000110       ; the number 0 )
WRITE 2001, %10111110       ; the number 1 )
WRITE 2002, %01001100       ; the number 2 )
WRITE 2003, %00011100       ; the number 3 ) for our particular wiring
WRITE 2004, %10110100       ; the number 4 ) setup this defines the
WRITE 2005, %00010101       ; the number 5 ) ten digits that we need.
WRITE 2006, %00000101       ; the number 6 )
WRITE 2007, %10011110       ; the number 7 )
WRITE 2008, %00000100       ; the number 8 )
WRITE 2009, %00010100       ; the number 9 )
                            ;
TRISA=%00010000             ; set up PORTA
TRISB=%00000000             ; set up PORTB
                            ;
P VAR BYTE                  ; pause variable
X VAR BYTE                  ; counter variable
Y VAR BYTE                  ; counter variable
Z VAR BYTE                  ; counter variable
Q VAR BYTE                  ; counter variable
PULSE_LEN VAR WORD          ; pulse length
VALUE VAR WORD              ; value of metronome counts
TOTAL VAR WORD              ; value of metronome counts
DIGIT1 VAR BYTE             ; each digit in display
DIGIT2 VAR BYTE             ; each digit in display
DIGIT3 VAR BYTE             ; each digit in display
DIGIT4 VAR BYTE             ; each digit in display
                            ;
X =0                        ; sets the initial value for X
P =400                      ; set pause will be in microseconds
TOTAL=0                     ; Initialize
Q =0                        ; initialize
                            ;
MAIN:                       ; main loop of program
  DIGIT1=-1                 ; set all 4 digits to a blank
  DIGIT2=-1                 ; set all 4 digits to a blank
  DIGIT3=-1                 ; set all 4 digits to a blank
  DIGIT4=-1                 ; set all 4 digits to a blank
```

(Continued)

Program 15.3 Finished pulse counting program (*Continued*)

```
  Q=Q+1                    ; counter to take jitter out of display
  IF Q<15 THEN             ; don't do anything
  ELSE                     ;
    PULSIN PORTA.4, 1, PULSE_LEN  ; read the pulse length on rise
    Q=0                    ; reset counter
  ENDIF                    ;
  VALUE=48500/PULSE_LEN    ; converts pulse length to count
  IF VALUE>1000 THEN DIGIT1=2000 + VALUE/1000       ; Separate
  IF VALUE>100 THEN DIGIT2=2000 + (VALUE//1000)/100 ; out the
  IF VALUE>10 THEN DIGIT3=2000 + ((VALUE//1000)//100)/10  ; four
  IF VALUE>0 THEN DIGIT4=2000 + ((VALUE//1000)//100)//10  ; digits
  GOSUB DISPLAY            ; show value on the 4 seven seg displays
GOTO MAIN                  ; do it over
                           ;
DISPLAY:                   ; reads each digit and then
  READ DIGIT1, VALUE       ; displays it
  PORTA=%00000001          ;
  GOSUB SHOW               ;
  PORTA=%00000010          ;
  READ DIGIT2, VALUE       ;
  GOSUB SHOW               ;
  PORTA=%00001000          ;
  READ DIGIT3, VALUE       ;
  GOSUB SHOW               ;
  PORTA=%00100000          ;
  READ DIGIT4, VALUE       ;
  GOSUB SHOW               ;
RETURN                     ;
                           ;
SHOW:                      ; shows each segment one at a time
  Z=%11111110              ; selects one segment at a time
  FOR X=1 TO 8             ; do the 8 segments, includes dec point
    Y=VALUE                ; get value to show
    Y.2=1                  ; inhibits the decimal point
    PORTB=VALUE | Z        ; makes value and z select one segment
    Z=(Z<<1)+1             ; go to next segment
    PAUSEUS P              ; pause to show segment
  NEXT X                   ; next segment
  PORTA=0                  ; clear PORTA
  PORTB=0                  ; clear PORTB to prevent ghosting
  PAUSEUS P                ; Pause to show clear
RETURN                     ; end of subroutine
                           ;
END                        ; end all programs with end
```

If we want to turn something on or off at a certain rpm with our controller, all we have to do is add the condition in the loop. For example, if we wanted to have an LED come on at 1750 rpm and go off at 1800 rpm, the code for adding that condition would be as follows. We will use the LED at PORTD.0.

Define the RPM variable and others with the rest of the variables. (See Program 15.4.)
Then after the LCDOUT line in the loop add the code segment in Program 15.4.

Program 15.4 Code segment for simple decision making

```
IF (RPM>1749) AND (RPM<1801) THEN   ;
  PORTD.0=1                         ;
ELSE                                ;
  PORTD.0=0                         ;
END IF                              ;
```

We now know how to read a frequency and display it either on the LCD or on a set of seven-segment displays.

Here (Program 15.5) is *another version* of that program that *actually counts how many pulses* are received in 0.25 seconds, and then multiplies the value by 4 and displays it four times a second. Since it takes the processor 0.25 seconds to count the pulses, you cannot avoid the flicker in a program as written in PBP.

Program 15.5 Code for using the 4 seven-segment displays

```
; The following are the images of the numbers
; stored in memory before we do anything. These will
; be used in PORTB to READ the number in each digit.

CLEAR                    ;
DEFINE OSC 4             ;
WRITE 1999, %11111111    ; blank, no LED is lit
WRITE 2000, %00000110    ; the number 0 in our case
WRITE 2001, %10111110    ; the number 1 in our case
WRITE 2002, %01001100    ; the number 2 in our case
WRITE 2003, %00011100    ; the number 3 in our case
WRITE 2004, %10110100    ; the number 4 in our case
WRITE 2005, %00010101    ; the number 5 in our case
WRITE 2006, %00000101    ; the number 6 in our case
WRITE 2007, %10011110    ; the number 7 in our case
WRITE 2008, %00000100    ; the number 8 in our case
WRITE 2009, %00010100    ; the number 9 in our case
                         ;
TRISA=%00010000          ; set up PORTA
TRISB=%00000000          ; set up PORTB
                         ;
P VAR BYTE               ; pause variable
X VAR BYTE               ; counter variable
Y VAR BYTE               ; counter variable
Z VAR BYTE               ; counter variable
Q VAR BYTE               ; counter variable
PULSES VAR WORD          ; pulse length
VALUE VAR WORD           ; value of metronome counts
TOTAL VAR WORD           ; value of metronome counts
```

(Continued)

Program 15.5 Code for using the 4 seven-segment displays (*Continued*)

```
DIGIT1 VAR BYTE        ; each digit in display
DIGIT2 VAR BYTE        ; each digit in display
DIGIT3 VAR BYTE        ; each digit in display
DIGIT4 VAR BYTE        ; each digit in display
                       ;
X=0                    ; sets the initial value for X
P=400                  ; set pause will be in microseconds
TOTAL=0                ; Initialize
Q=0                    ; initialize
                       ;
MAIN:                  ; main loop of program
  DIGIT1=-1            ; set all 4 digits to a blank
  DIGIT2=-1            ; set all 4 digits to a blank
  DIGIT3=-1            ; set all 4 digits to a blank
  DIGIT4=-1            ; set all 4 digits to a blank
  Q=Q+1                ; counter to take jitter out of display
  IF Q<15 THEN         ; don't do anything
  ELSE                 ;
    COUNT PORTA.4, 250, PULSES  ; read the pulses
    Q=0                           ; reset counter
  ENDIF                ;
  VALUE=4*PULSES       ; converts pulses to count
  IF VALUE>1000 THEN DIGIT1= 2000 + VALUE/1000  ; Separate
  IF VALUE>100 THEN DIGIT2= 2000 + (VALUE//1000)/100  ; out the
  IF VALUE>10 THEN DIGIT3= 2000 + ((VALUE//1000)//100)/10; four
  IF VALUE>0 THEN DIGIT4= 2000 + ((VALUE//1000)//100)//10; digits
  GOSUB DISPLAY        ; show value on the 4 seven seg. displays
GOTO MAIN              ; do it over
                       ;
DISPLAY:               ; reads each digit and then
  READ DIGIT1, VALUE   ; displays it
  PORTA=%00000001      ;
  GOSUB SHOW           ;
  PORTA=%00000010      ;
  READ DIGIT2, VALUE   ;
  GOSUB SHOW           ;
  PORTA=%00001000      ;
  READ DIGIT3, VALUE   ;
  GOSUB SHOW           ;
  PORTA=%00100000      ;
  READ DIGIT4, VALUE   ;
  GOSUB SHOW           ;
RETURN                 ;
                       ;
SHOW:                  ; shows each segment one at a time
  Z=%11111110          ; selects one segment at a time
  FOR X=1 TO 8         ; do the 8 segments, includes dec. point
```

(*Continued*)

Program 15.5 Code for using the 4 seven-segment displays (*Continued*)

```
            Y=VALUE              ; get value to show
            Y.2=1                ; inhibits the decimal point
            PORTB=VALUE | Z      ; makes value and z select one segment
            Z=(Z<<1)+1           ; go to next segment
            PAUSEUS P            ; pause to show segment
        NEXT X                   ; next segment
    PORTA=0                      ; clear PORTA
    PORTB=0                      ; clear PORTB to prevent ghosting
    PAUSEUS P                    ; pause to show clear
    RETURN                       ; end of subroutine
                                 ;
    END                          ; end all programs with end
```

Here is another way in which I made the tachometer after I needed a board that could be used as a base for all kinds of devices. I designed this board as a universal all-purpose board for my experiments. I have since improved the board and have been very happy with the results. It saved me much time and sped up the many experiments I had to make to write this book.

Figures 15.12 and 15.13 show the use of a flexible board that supports the use of both 4 seven-segment displays and that of a 2-line-by-20-character display. All the wiring to both types of devices is in place. This eliminates most of the tedious wiring that has to be done on almost all projects. On these cards, all the I/O has been left unconnected

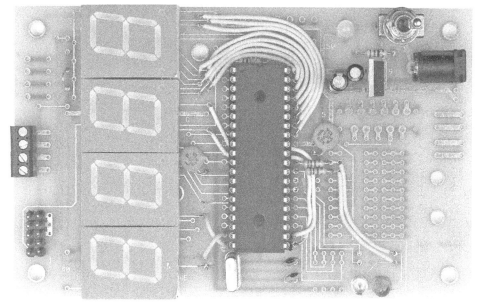

Figure 15.12 Another way to make the tachometer/pulse counter, front.

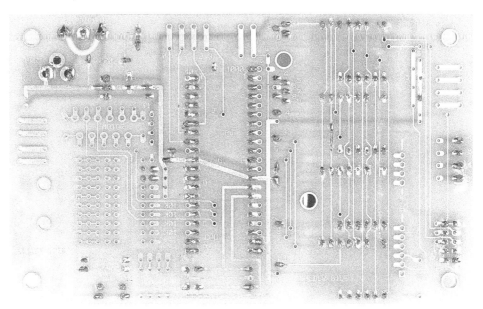

Figure 15.13 **The back of the tachometer/pulse counter.** (Note how the custom wiring connects up the I/O only. The rest is pre-wired.)

and has to be wired the way you see fit. On the other hand, the power supply and the basic connections to the CPU are already connected and ready to use. If you plan to build many different instruments and controllers, you should consider designing a board based on this design. If you need just one or a few boards, these boards are available at encodergeek.com.

16

CREATING ACCURATE INTERVALS WITH TIMERS: THE METRONOMES

Project 2

This chapter is about learning to use the timers in a PIC.

First a note about the timer functions and availability. Though there are four timers in the PIC 16F877A, not all the timers are available for our projects, nor are they available at all times. The other uses for which a timer may be employed are:

- Used with the HPWM instruction
- Used for communications
- Control of the slave port
- Baud rate generation
- Resources are shared with a watchdog timer (Watchdog timer [timer #4] is not available for any programming purposes but does share scaling resources with Timer2.)

GETTING COMFORTABLE

Our goal is to be comfortable using the three timers after we have created the metronomes for this project. All the metronomes created in software and hardware behave identically but use different timers.

The metronome project is essentially a software project in that it is about learning how to use the timers. We should already have the two pieces of hardware needed to run the metronome software. They are, of course, the LAB-X1 and the board we made for the last project, the tachometer (to which we will have to add connection points for a potentiometer and a speaker [or LED]). The existing piezo beeper on the LAB-X1 may also be used.

Figure 16.2 show that back of the metronome board shown in Figure 16-1.

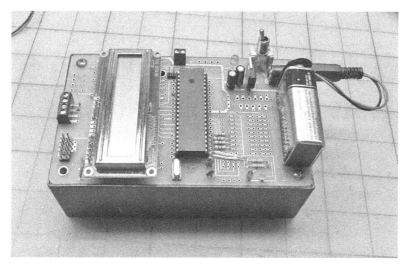

Figure 16.1 The electronic metronome: the card fits nicely on a box from All Electronics. (Uses the same card as that for the second tachometer.)

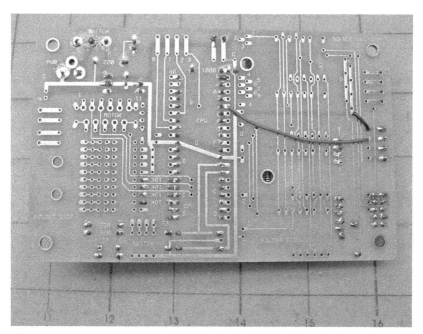

Figure 16.2 Electronic metronome back—Custom-made card. (Some of the wiring is on the back of the card to provide a flexible wiring system. Most I/O points are not attached to anything and can therefore be connected to any of the MCU pins with suitable jumpers.)

We will use each of the three timers (we do not have the ability to use the watchdog timer) in the PIC 16F877A to create instruments that provide us with accurate predictable timed intervals we can select with a potentiometer. Two of the three timers can also be used as counters, and that will be undertaken in the next chapter where we count marbles in some of the many different ways available to us. Counters are essentially timers that get their count pulse from an external source.

On the metronomes created, the counts per minute will be controlled by a potentiometer, they will be displayed on the display, and they will be annunciated on a speaker (or LED). Since both the tachometer hardware and the LAB-X1 hardware will have these accessories on them, either one can be programmed for the purpose we have in mind. Though the output needed from a metronome is rather limited (40 to 208 ticks per minute is the standard), our project will be able to provide any count desired, from a count every few seconds to a few thousand counts per second. All we will need to do is change a few lines of code in the software. That, in a nutshell, is the power of instruments that you yourself make.

The three timers in the PIC 16F877A have the following salient features, and the characteristics of the timers in most other PICs are similar. Once you get comfortable with the three timers in this PIC, you should have no difficulty with any timer in any PIC. The hard part will be getting familiar with the new datasheet for the PIC you select.

TIMER0

- This is an 8-bit timer.
- Official name TMR0.
- Register used TMR0.
- This register can be written to and read from at will.
- *Runs all the time*, cannot be stopped or started.
- Uses a prescalar.
- Does not use a postscalar.
- Can run from an internal clock (at Fosc/4).
- Can be used with an external signal for counting.
- Generates a programmable interrupt on overflow from 255 to 0.

TIMER1

- This is a 16-bit timer.
- Official name: TMR1.
- Registers used: TMR1H and TMR1L for the high and low bytes of the 16-bit word.
- The two registers mentioned in the previous bullet point can be written to and read from.
- The timer can be turned on and off.
- Uses a prescalar.
- *Does not* use a postscalar.
- Can run from an internal clock (at Fosc/4).
- Can also run from an external clock for timing *and* counting.
- Generates a programmable interrupt.

TIMER2

- This is an 8-bit timer.
- Official name: TMR2.
- Register used: TMR2.
- This timer can be written to and read from.
- Has a register it can be compared to in order to generate an interrupt.
- The special 8-bit register that is compared to the timer can be written to. An interrupt is generated when the comparison yields the specified match.
- The timer can be turned on and off.
- Uses a prescalar.
- Uses a postscalar.
- Cannot use an external clock and so *cannot be used as a counter*.
- Generates a programmable interrupt.
- Shares certain functions with the watchdog timer. This is important.
- This timer has a special function that is used to drive a synchronous port for transferring data between the PIC and a computer.

DEFINING SOME TERMS

Before we start using the timers, let's define a few timer-related words and concepts that will help us understand the functions of the timers more easily.

HPWM Commands When using the HPWM commands, a timer must be assigned to help generate the frequencies used by these commands. The timer assigned to this function is then not available for any other use. In general, each timer can be used to support one function at one time and is busy to all other functions. (The default timer for HPWM use is Timer1, the 16-bit timer).

Prescalar A prescalar is a divider function that slows down the rate at which the interrupts are generated by the timer. If a prescaler is set to 8, the timer clock feed will be divided by eight times before if is fed to the timer count. This slows down the rate at which the interrupts occur by eight. Prescalars are specified by setting a few bits in one of the registers that control the timer. The registers and prescalar values are different for each timer.

Postscalar A postscalar has the same effect as a prescalar, it too slows down the rate at which the interrupts occur, but instead of slowing down the clock feed, it counts the overflows. If a postscalar is set for 4, the timer will overflow four times before an interrupt is generated.

The prescalar and postscalar values are multiplied by one another to get the total delay seen on the interrupts.

Interrupt An interrupt is a signal that informs the system that something needs immediate attention. The interrupt manifests itself by setting a bit somewhere in the processor.

The compiler software for the system is designed to be able to jump to a designated subroutine when this happens. We specify the target destination for this with the ON INTERRUPT GOTO command.

Interrupt Flag An interrupt flag is a bit that is set in one of the registers associated with a timer when the timer overflows and after all the pre- and postscalar effects have been taken into account. It indicates that the program is ready to be given attention by the main part of the processor. After the interrupt has been taken care of, the interrupt bit must be set to 0 (cleared) by the interrupt service routine to allow it to be set to 1 again when the next interrupt takes place. The bit is usually cleared at the end of the interrupt-handling routine. The interrupt-handling routine must be short enough to complete before the next interrupt arrives. If this is not the case, an interrupt will be lost. (This is considered fatal and is unacceptable.)

Interrupt Enable Bit An interrupt enable bit is the bit that has to be set to 1, within one of the registers that controls the timer, before an interrupt can be generated. This bit can usually be turned on and off by the program. Before a specific interrupt can take place, its specific enable bit must be set to 1.

Interrupt Latency Beware that the interrupts are not handled as soon as they occur. The PBP compiler inhibits all interrupts while it executes an instruction. If an interrupt occurs while an instruction is being executed, the interrupt will be addressed after the instruction is complete. This can lead to a particularly long delay in the case of a long PAUSE instruction. PAUSE instructions should be broken up into smaller pauses in loops to reduce this latency so the largest latency encountered can be tolerated by the task at hand. Though interrupts may occur exactly and as frequently as programmed, they may not be handled immediately by the program. The programmer needs to be aware of this condition when programming for time-critical applications. (Assembly language programming does not have the problem to this extent. It depends on how the assembler is designed, but even so it takes some time to get the interrupt-handling task accomplished.)

Global Interrupt Enable (GIE) Bit The GIE bit in the PIC allows you to shut off all interrupts when it is set to 0. It is located at INTCON.7. On two of the timers—Timer1 and Timer2—this global interrupt enable bit must be enabled before any interrupt can occur. Timer0 runs all the time and is not affected by the global interrupt enable bit. See datasheet page 22.

Whenever any interrupt flag is set, the GIE bit is turned off by the operating system. When the flag is cleared by you (with the software), the GIE bit will be reset automatically by the operating system. This means no other interrupt can occur until you take care of the interrupt that has occurred. If you are using more than one interrupt, this can cause problems, and addressing these problems is beyond the scope of this book. Beginners are advised to use only one interrupt in their programs till they get proficient enough with the PICs to create more complicated programs.

(For those who have an insatiable need to try these things out, the solution to the previously mentioned problem has to do with replacing the interrupts very quickly with internal flags, which are addressed later as a part of the main loop, and then proceeding with the program. So the interrupt routine for each interrupt just sets its own designated flag, resets its interrupt flag and ends. When the main loop sees this designated flag, it takes care of it as part of the main routine, not the interrupt routine. In the mean time, you can address the next interrupt flag. This gets complicated in a hurry and the program designer must guarantee that doing it this way will actually work. In other words, this is not trivial, so it is best avoided for now.)

Timers are not the only things that create interrupts. External events can be used to initiate an interrupt. PORTB has special significance in the generation of these interrupts. See the datasheet.

Now let's look at each timer in greater detail.

Timer0

Timer0 is discussed on page 51 in the datasheet.

Basic description: Timer0 has 8 bits, runs all the time, and sets its interrupt bit every 256 counts of Fosc/4 if the bit has been cleared. It can use an internal or external signal (for counter use). It can count on a falling or rising edge. The frequency of the interrupt can be modified with a prescalar, and the register itself can be read and written to.

The two registers that control Timer0 are OPTION_REG and INTCON. The bits in these registers affect Timer0 as follows:

OPTION REGISTER values

OPTION_REG.0	Bit0	)
OPTION_REG.1	Bit1	) These 4 bits control the value
OPTION_REG.2	Bit2	) of the Prescalar as listed below
OPTION_REG.3	Bit3	)
OPTION_REG.4	Bit4	Selects high to low transition of clock when set to 1 Selects low to high transition of clock when set to 0
OPTION_REG.5	Bit5	Selects internal clock for timer when set to 0 Selects external clock on TOCK1* for counter when set to 1
~~OPTION_REG.6~~	~~Bit6~~	~~Not used to control Timer0.~~
~~OPTION_REG.7~~	~~Bit7~~	~~Not used to control Timer0. PortB pullups.~~

*TOCK1 is pin 6 on the PIC 16F877A, also known as PORTA.4.
PORTA Pin 4 is the fifth pin on *PORTA*.

4 Bit value	Prescalar value	
0000	1:2	
0001	1:4	
0010	1:8	
0011	1:16	
0100	1:32	
0101	1:64	
0110	1:128	
0111	1:256	
1xxx	1:1	No prescalar applied if bit 3 is set to 1. Prescalar is therefore OFF.

INTCON is the interrupt control register

~~INTCON 0~~	~~Bit0~~	~~Not used to control Timer0~~
~~INTCON 1~~	~~Bit1~~	~~Not used to control Timer0~~
INTCON 2	Bit2	Interrupt flag
~~INTCON 3~~	~~Bit3~~	~~Not used to control Timer0~~
~~INTCON 4~~	~~Bit4~~	~~Not used to control Timer0~~
INTCON 5	Bit5	Interrupt enable bit
INTCON 6	Bit6	Used with Interrupt enable bit
~~INTCON 7~~	~~Bit7~~	~~Not used to control Timer0~~

As mentioned in the preceding, there is no way to start and stop Timer0. It runs all the time. However, since it can be written to, it can be set to 0 or any value up to 255 whenever desired. Also in the counter mode, if there is no signal on TOCK1, the counter will not increment, and so no interrupt will occur—thus, in a sense the counter function (but not the internally clocked timer function) can be stopped. If running free, Timer0 generates an interrupt every 256 Fosc/4 cycles whether any attention is paid to the interrupt or not, so if its interrupt flag is never cleared, it stays set.

Watchdog timer interaction: If the prescalar is being used by the watchdog timer, it cannot be used by Timer2 and visa versa.

Now that we know where all the relevant bits are, we can write a program for the LAB-X1 to test things out. In this program, we will set the prescalar to its maximum value to slow things down as far as possible so we can see what is going on more easily.

Here is the plan for testing the operation of Timer0:

1. We will increment the value of the variable X in the interrupt routine. Therefore, we will be sure we have entered and returned from the interrupt routine if this value is being incremented.
2. We will display the value of X in the main loop. Therefore, if we see X incremented, the interrupts are being called and returned from while we are in the main loop.
3. If we see the preceding two things taking place, we will have successfully used Timer0. It's that simple. We will be ready to use Timer0 in our programs.

These first programs for Timer0 are heavily commented, so you can see exactly what is being done. The programs that follow are less heavily commented because they are very similar to these first programs.

First, let's set up the LCD the usual way, as shown in Program 16.1.

Program 16.1 Basic interrupt routine for Timer1

```
CLEAR                         ; always start with clear
DEFINE OSC 4                  ; define oscillator speed
DEFINE LCD_DREG PORTD         ; define LCD connections
DEFINE LCD_DBIT 4             ; 4 bit path
DEFINE LCD_RSREG PORTE        ; select reg
DEFINE LCD_RSBIT 0            ; select bit
DEFINE LCD_EREG PORTE         ; enable register
DEFINE LCD_EBIT 1             ; enable bit
LOW PORTE.2                   ; make low for write only
                              ;
; Set the port directions. We must set all of PORTD and all of
; PORTE as outputs
; even though PORTE has only 3 lines. The other 5 lines will be
; ignored by the system.
; PORTC is needed because the piezo speaker is on pin 2 of this
; port.
                              ;
TRISC = %11111001             ; set PORTC.1 and PORTC.2 as output
                              ; for the speaker and LED connections
TRISD = %00000000             ; set all PORTD lines to output
TRISE = %00000000             ; set all PORTE lines to output
X VAR WORD                    ; set up the variable
ADCON1=%00000111              ; set the Analog-to-Digital control
                              ; register
                              ; needed for the 16F877A see notes
PAUSE 500                     ; pause for LCD to start up
LCDOUT $FE, 1                 ; clear screen
                              ;
ON INTERRUPT GOTO INT_ROUTINE ; tells program where to go on an
                              ; interrupt
INTCON.5=1                    ; sets up the interrupt enable
```

(Continued)

Program 16.1 Basic interrupt routine for Timer1 (*Continued*)

```
INTCON.2=0                    ; clears the interrupt flag so it
                              ; can be set
OPTION_REG=%00000111          ; sets the prescalar to 256
X=0                           ; sets the initial value for X
LCDOUT $FE, $80, "Metronome"  ; display first line
                              ;
MAIN:                         ; the main loop of the program
  LCDOUT $FE, $C0, DEC5 X     ; write X to line 2
  IF X>=15 THEN               ; check value of X
    X=0                       ; reset the value
    TOGGLE PORTC.2            ; toggle the speaker so we can hear it
    TOGGLE PORTC.1            ; for the ext speaker or an LED
  ENDIF                       ; end of testing X
GOTO MAIN                     ; repeat loop
                              ;
DISABLE                       ; reqd instruction, to the compiler
INT_ROUTINE:                  ; interrupt service routine
  X=X+1                       ; Increment the counter
  INTCON.2=0                  ; Clear the interrupt flag
RESUME                        ; Go back to where you were
ENABLE                        ; reqd instruction, to the compiler
                              ;
END                           ; all programs must end with End
```

This program toggles line C.2 each time through its cycle and gives us a click about once a second, or once every 15 interrupts, using a prescalar of 256. This means that the fastest interrupts we could get with Timer0 would come 256 × 15 faster, or at about 3840 interrupts per second if we did not change the clock rate or try to load the timer any other way. The number 3480 is good to remember when using Timer0 because it defines the empirical maximum interrupt rate for us.

- 4,000,000/4/256 = 3,906.25 (the theoretical value for a 4 MHz processor)
- 3906.25/256 = 15.26 (used 15 in the preceding program since it is the closest integer)
- 15*256 = 3480 (as mentioned earlier, not an exact value)

The prescalar value we use has an important effect on the length of the interrupt service routine. If we are using a small value for the prescalar, the time available between interrupts becomes very small and the moment may come when all the time available is used up servicing the interrupt routine, leaving no time to do the foreground task. We will investigate this by setting the prescalar with the first potentiometer on the LAB-X1 board and loading its value into the prescalar in real time to see what happens. To do this, we need the following code added to the program. It consists of reading the potentiometer and then placing the read value into the four prescalar bits.

The code for reading POT 1 is shown in Program 16.2.

Program 16.2 Basic program to see the effect of the prescalar value on Timer0 operation

```
A2D_V VAR    BYTE              ; create A2D_Value to store result
DEFINE ADC_BITS 8              ; set number of bits in result
DEFINE ADC_CLOCK 3             ; set clock source (3=rc)
DEFINE ADC_SAMPLEUS 50         ; set sampling time in uS

; The value is read with
ADCIN 0, A2D_V                 ; read channel 0 to A2D_V

; The value is displayed on the LCD to three places with
LCDOUT $FE,  $C0,  DEC3 A2D_V

; The value is then placed in the prescalar with
OPTION_REG= (A2D_V / 32)
```

When these lines of code are added to the program the program becomes:

```
CLEAR                          ; always start with clear
DEFINE OSC 4                   ; define oscillator speed
DEFINE LCD_DREG PORTD          ; define LCD connections
DEFINE LCD_DBIT 4              ; 4 bit path
DEFINE LCD_RSREG PORTE         ; select reg
DEFINE LCD_RSBIT 0             ; select bit
DEFINE LCD_EREG PORTE          ; enable register
DEFINE LCD_EBIT 1              ; enable bit
DEFINE ADC_BITS 8              ; set number of bits in result
DEFINE ADC_CLOCK 3             ; set clock source (3=rc)
DEFINE ADC_SAMPLEUS 50         ; set sampling time in uS
LOW PORTE.2                    ; make low for write only
                               ;
; Set the port directions. We must set all of PORTD and all of
; PORTE as outputs
; even though PORTE has only 3 lines. The other 5 lines will
; be ignored by the system.
; PORTC is needed because the piezo speaker is on pin 2 of this port.
                               ;
TRISC = %11111011              ; set port c.2 to speaker
                               ; connection to output
TRISD = %00000000              ; set all PORTD lines to output
TRISE = %00000000              ; set all PORTE lines to output
X VAR WORD                     ; set up the variable
A2D_V VAR BYTE                 ; create A2D_Value to store result
ADCON1=%00000111               ; Set the Analog-to-Digital control
                               ; register
                               ; needed for the 16F877A see notes
PAUSE 500                      ; pause for LCD to start up
LCDOUT $FE, 1                  ; clear screen
ON INTERRUPT GOTO INT_ROUTINE  ; tells program where to go on
                               ; interrupt
```

(Continued)

Program 16.2 Basic program to see the effect of the prescalar value on Timer0 operation (Continued)

```
INTCON.5=1                        ; sets up the interrupt enable bit
INTCON.2=0                        ; clears the interrupt flag so it can
                                  ; be set
X=0                               ; sets the initial value for X
LCDOUT $FE, $80, "Metronome"      ; display first line
                                  ;
MAIN:                             ; the main loop of the program
  ADCIN 0, A2D_V                  ; read channel 0 to A2D_Value
  OPTION_REG= (A2D_V / 32)        ; set the option register low nibble
  LCDOUT $FE,  $C0,  DEC3 A2D_V/32,"  ; display value
GOTO MAIN                         ; do it again
                                  ;
DISABLE                           ; reqd Instruction to the compiler
INT_ROUTINE:                      ; interrupt service routine
  TOGGLE PORTC.2                  ; toggle the port
  X=X+1                           ; Increment the counter
  INTCON.2=0                      ; clear the interrupt flag
RESUME                            ; go back to where you were
ENABLE                            ; reqd Instruction to the compiler
                                  ;
END                               ; all programs must end with End
```

In Program 16.2, the LCD shows us the value of X and the bits that have been set in the OPTION_REG register. We have had to remove the comparison and resetting of the X variable because it takes too much time and the toggling has been moved into the interrupt service routine, so you can hear how often the interrupt is being called. Notice that as the interrupts become more frequent, the incrementing of the value of X in the main routine cannot keep up with the speed with which the interrupts are arriving. Even though a minimal amount of work is being done in the interrupt service routine, it is too much.

We can add to the time taken by the interrupt service routine by adding a PAUSE in the routine. Play with adding a PAUSE of between 1 and 25 μsec in the routine to see what happens. (Interrupts are missed and the toggling does not respond to the changes made to the option register bits.)

Now that we have a feel for the problems involved, let's write the metronome program. A standard metronome provides between 40 and 208 counts per minute. We are going to use POT 0 to control the rate so the 256 values that can be read for the POT must be mapped to the 168 rates (208 − 40) needed by the metronome, and each rate must be accurate enough to serve everyday musical needs.

The formula for converting from 0 to 255 to from 40 to 208 is

Ticks = 40 + [(208 − 40) * POT.0]/255

Looking back at 3480 as the maximum number of interrupts we can handle with ease, and seeing that we need to generate 168 different frequencies, we see that we can have about 3480/168 = 20 interrupt counts separating each frequency, so we know that they can all be differentiated. If we set the option register to %00000011, we will get a prescalar of 16 and increase the number of interrupts received by a factor of 16 (256/16). This will be enough to do the job. Increasing the number of interrupts by a factor of 16 means we need to look at 16 times more counts on the X counter before we toggle the speaker.

The program for a metronome using Timer0 and the LAB-X1 is shown in Program 16.3.

Program 16.3   Working metronome based on Timer0 and the LAB-X1

```
CLEAR                            ; always start with clear
DEFINE OSC 4                     ; define oscillator speed
DEFINE LCD_DREG PORTD            ; define LCD connections
DEFINE LCD_DBIT 4                ; 4 bit path
DEFINE LCD_RSREG PORTE           ; select reg
DEFINE LCD_RSBIT 0               ; select bit
DEFINE LCD_EREG PORTE            ; enable register
DEFINE LCD_EBIT 1                ; enable bit
DEFINE ADC_BITS 8                ; set number of bits in result
DEFINE ADC_CLOCK 3               ; set clock source (3=rc)
DEFINE ADC_SAMPLEUS 50           ; set sampling time in uS
LOW PORTE.2                      ; make low for write only
                                 ;
; Set the port directions. We must set all of PORTD and all of
; PORTE as outputs
; even though PORTE has only 3 lines. The other 5 lines will
; be ignored by the system.
; PORTC is needed because the piezo speaker is on pin 2 of this port.
                                 ;
TRISC = %11111011                ; set PORTC.2 to speaker connection
                                 ; to output
TRISD = %00000000                ; set all PORTD lines to output
TRISE = %00000000                ; set all PORTE lines to output
X VAR WORD                       ; set up the variable
POTPOS VAR WORD                  ;
A2D_V VAR BYTE                   ; create A2D_Value to store result
ADCON1=%00000111                 ; set the Analog-to-Digital
                                 ; control register
                                 ; needed for the 16F877A see notes
PAUSE 500                        ; pause for LCD to start up
LCDOUT $FE, 1                    ; clear screen
ON INTERRUPT GOTO INT_ROUTINE    ; tells program where to go on
                                 ; interrupt
INTCON.5=1                       ; sets us the interrupt enable
INTCON.2=0                       ; clears the interrupt flag so it
                                 ; can be set
X=0                              ; sets the initial value for X
                                 ;
MAIN:                            ; the main loop of the program
  ADCIN 0, A2D_V                 ; read channel 0 to A2D_Value
  OPTION_REG=%00000011           ; prescalar is 16
```

(Continued)

Program 16.3 Working metronome based on Timer0 and the LAB-X1 (*Continued*)

```
    POTPOS= 208-(((208-40)*A2D_V)/255)   ; the potentiometer
                                          ; position value
    LCDOUT $FE, $80, DEC3 POTPOS,"  "    ; Display the position
    LCDOUT $FE,  $C0, DEC X ,"  "        ; display the X count
    IF X>=16*15*60/POTPOS THEN           ; this is where we add the multiply
                                          ; by 16
      TOGGLE PORTC.2                     ; toggle the speaker
      X=0                                ; reset x
    ELSE                                 ;
    ENDIF                                ;
GOTO MAIN                                ; go back and do it again
                                          ;
DISABLE                                  ; required by the compiler
  INT_ROUTINE:                           ; interrupt service routing
    X=X+1                                ; increment value of X
    INTCON.2=0                           ; reset/clear the flag
  RESUME                                 ; go back to interrupt point
ENABLE                                   ; required by the compiler
END                                      ; end all programs with end
```

In the wiring diagram shown in Figure 16.3, a ten-pin connector is shown (for programming the PIC in place) on the circuit board. This is an important convenience that should not be omitted on any programmable device you create. You will use it many, many times as you experiment with the project, and not having to take the PIC out of its socket to reprogram it will save you a lot of headaches. Take the time to wire in the programmer connection (in all your projects).

Next, we need to convert the program we have developed to run on the tachometer hardware with its seven-segment displays. To allow us to turn the tachometer into a metronome, we also have to make two hardware additions. We need to add a potentiometer and a speaker to the card. The circuitry for adding these two items is shown circled in Figure 16.3.

In order to display on the 4 seven-segment LED displays on the tachometer, we must add the code for the displays in place of the instructions for the LCD. This code is taken from the code in the tachometer program (with minor modifications as needed) shown in Program 16.4.

Program 16.4 Program to use the tachometer board as a metronome

```
CLEAR                      ; clear memory
DEFINE OSC 4               ; osc speed
; The following are the images of the numbers
; stored in memory before we do anything. These will
; be used in PORTB to set the number in each digit.
; Your wiring will require numbers to suit it.
;
WRITE 1999, %11111111      ; blank, no LED is lit
```

(*Continued*)

Program 16.4 Program to use the tachometer board as a metronome
(*Continued*)

```
WRITE 2000, %00000110    ; the number 0   )
WRITE 2001, %10111110    ; the number 1   )
WRITE 2002, %01001100    ; the number 2   )
WRITE 2003, %00011100    ; the number 3   ) will depend on
WRITE 2004, %10110100    ; the number 4   ) how your particular
WRITE 2005, %00010101    ; the number 5   ) engine is wired.
WRITE 2006, %00000101    ; the number 6   )
WRITE 2007, %10011110    ; the number 7   )
WRITE 2008, %00000100    ; the number 8   )
WRITE 2009, %00010100    ; the number 9   )
                         ;
DEFINE ADC_BITS 8        ; set number of bits in result
DEFINE ADC_CLOCK 3       ; set clock source (3=rc)
DEFINE ADC_SAMPLEUS 50   ; set sampling time in uS
                         ;
TRISA=%00000000          ; set up PORTA
TRISB=%00000000          ; set up PORTB
TRISC=%11111001          ; set up PORTC
TRISE=%00000100          ; set up PORTE
P VAR WORD               ; pause variable
X VAR BYTE               ; counter variable
NUM VAR WORD             ; number read from Pot
LEDPOS VAR WORD          ; position in display
VALUE VAR WORD           ; value of metronome counts
DIGIT VAR BYTE           ; each digit in display
P=4                      ; pause in microseconds
ADCON1=%00000000         ; this selects first line in the table
INTCON.5=1               ; sets us the interrupt enable
INTCON.2=0               ; clears the interrupt flag so it can be set
OPTION_REG=%00000011     ; sets the prescalar to 16
X=0                      ; sets the initial value for X
HIGH PORTC.2             ; sets pin high to start with
ON INTERRUPT GOTO INT_ROUTINE  ; tells program where to go on
                               ; interrupt
                         ;
MAIN:                    ; main loop of program
  ADCIN 7, NUM           ; read channel 0 pot value
  LEDPOS=208-(((208-40)*NUM)/255)    ; the potentiometer
                                     ; position value
  VALUE=LEDPOS                       ; remember value we are
                                     ; working with
  IF X>=16*15*30/LEDPOS THEN  ; this is where we add the
                              ; multiply by 16
    TOGGLE PORTC.2       ; toggle the speaker
    X=0                  ; reset x
  ELSE                   ;
  ENDIF                  ;
```

(*Continued*)

Program 16.4 Program to use the tachometer board as a metronome
(*Continued*)

```
  GOSUB DISPLAY              ; show value on the 4 seven seg displays
GOTO MAIN                    ; do it again forever
                             ;
DISPLAY:                     ; the routine to display the 4 numbers
  IF VALUE<1000 THEN         ; check value for first digit
    DIGIT=1999               ; set a blank
  ELSE                       ;
    DIGIT=LEDPOS/1000 +2000    ; set pattern position in table
  ENDIF                      ;
  READ DIGIT, PORTB          ; read the pattern for the digit
  PORTA=%00000001            ; display digit 1 from if 1000 to 9999
  PAUSE P                    ; pause to let displays come ON
                             ;
  LEDPOS=LEDPOS//1000        ; get Remainder
  IF VALUE<100 THEN          ; check value for second digit
    DIGIT=1999               ;
  ELSE                       ;
    DIGIT=LEDPOS/100 +2000   ;
  ENDIF                      ;
  READ DIGIT, PORTB          ;
  PORTA=%00000010            ; display digit 2 from 100 to 999
  PAUSE P                    ;
                             ;
  LEDPOS=LEDPOS//100         ;
  IF VALUE<10 THEN           ; check value for third digit
    DIGIT=1999               ;
  ELSE                       ;
    DIGIT=LEDPOS/10 +2000    ;
  ENDIF                      ;
  READ DIGIT, PORTB          ;
  PORTA=%00001000            ; display digit 3 from 10 to 99
  PAUSE P                    ;
                             ;
  LEDPOS=LEDPOS//10          ;
  DIGIT=LEDPOS +2000         ;
  READ DIGIT, PORTB          ;
  PORTA=%00100000            ; display last digit form 0 to 9
  PAUSE P                    ;
RETURN                       ; end of routine
                             ;
DISABLE                      ; required by the compiler
INT_ROUTINE:                 ; interrupt service routing
  X=X+1                      ; increment value of X
  INTCON.2=0                 ; reset/clear the flag
RESUME                       ; go back to interrupt point
ENABLE                       ; required by the compiler
                             ;
END                          ; end all programs with end
```

Note *This program will be modified for each of the timers, but since it does not really change when the other timers are used, I will not list this program again. The changes to the timers will be the same as will be made for the LCD on the LAB-X1. These changes are listed under each timer section.*

Timer1

Timer1 is discussed on page 55 in the datasheet.

Basic description: Timer1 is a 16-bit timer with a prescalar. It can be turned on and off, can use an internal or external crystal, can be synchronized with an internal or external clock, and be used as a counter.

Let's see what it takes to create the preceding metronome with Timer1, what must be changed, and what the problems and advantages of using this particular timer are.

Timer1 is the only 16-bit timer in the 16F877A. The longest interval that can be timed with a 4 MHz clock is a little over half a second. This is calculated as follows.

The maximum prescale value is 1:8

Instruction clock cycle (1 uS @ 4 MHz)

Counts from one overflow to the next (65536)

$$1 \text{ uS} * 8 * 65536 = 0.524288 \text{ seconds}$$

This timer uses two bytes, identified as TMR1H and TMR1L (the high and low byte). These 2 bytes cannot be read or written in one instruction, and the attendant problems having to do with the low byte overflow or underflow affecting the high byte in the middle of a read or write. It is a problem that must be taken care of when reading and writing to these 2 bytes. (The usual advice is to stop the timer before reading or writing to it, and then start it again immediately after that is done.)

The registers that control Timer1 are *T1CON, INTCON, PIE1,* and *PIR1*. The bits in these registers are used to control Timer1 as shown next:

Bits T1CON.5 and T1CON.4 set the value of the prescalar as follows. This timer does not have a postscalar capability.

Bit value		Prescalar value selected
T1CON.5, 4 =00	1	meaning that no prescalar is used
T1CON.5, 4 =01	2	
T1CON.5, 4 =10	4	
T1CON.5, 4 =11	8	

PIE1.0 and INTCON.6 control the interrupt enable bit; both must be set to 1
The interrupt flag is located at PIR1.0 and must be cleared in the interrupt routine.

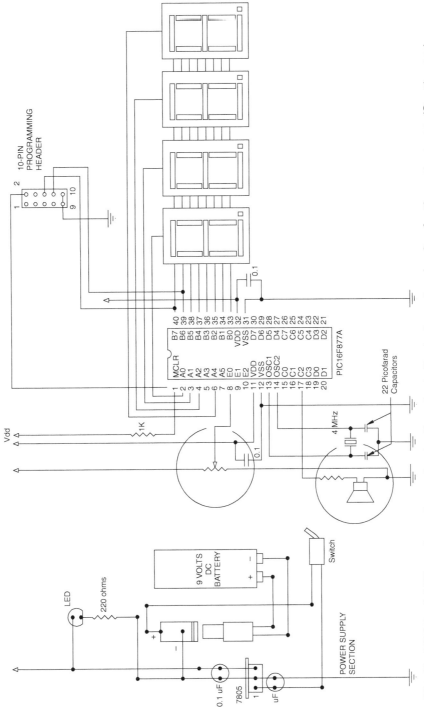

Figure 16.3 Addition of speaker and potentiometer to the tachometer circuitry for the metronome. (Speaker and potentiometer have been added.)

Like all the timers, Timer1 is clocked by the crystal frequency divided by 4 (or Fosc/4). It can also be clocked by an external oscillator connected between PORTC.0 and PORTC.1. The selection between the internal and external oscillators is made by what is in T1CON.1 and T1CON.3, as follows:

T1CON.1=0   Selects the Fosc/4 signal

T1CON.1=1   Selects the external input between PORTC.0 and PORTC.1

T1CON.3=1   Also must be set if T1CON.1 is set

The use of *T1CON* to control Timer1 is summarized as follows:

T1CON.7   ~~Not used.~~

T1CON.6   ~~Not used.~~

T1CON.5   ) These 2 bits

T1CON.4   ) specify the prescalar

T1CON.3   Set to 1 to enable oscillator

T1CON.2   Set to 1 to selects external clock

T1CON.1   Selects external clock when set to 1, internal clock when 0

T1CON.0   Turns Timer1 on when set to 1, off when 0, becomes counter

USING TIMER1

The listing in Program 16.5 shows how Timer1 can be used to create a metronome. Compare this to the listing for Timer0 mentioned previously.

Program 16.5   Metronome using Timer1

```
CLEAR                         ; clear
DEFINE OSC 4                  ; define oscillator speed
DEFINE LCD_DREG PORTD         ; define LCD connections
DEFINE LCD_DBIT 4             ; 4 bit path
DEFINE LCD_RSREG PORTE        ; select reg
DEFINE LCD_RSBIT 0            ; select bit
DEFINE LCD_EREG PORTE         ; enable register
DEFINE LCD_EBIT 1             ; enable bit
LOW PORTE.2                   ; make low for write only
```

(*Continued*)

Program 16.5 Metronome using Timer1 (*Continued*)

```
; Set the port directions. We must set all of PORTD and all of
; PORTE as outputs
; even though PORTE has only 3 lines. The other 5 lines will
; be ignored by the system.
; PORTC is needed because the piezo speaker is on pin 2 of this
; port.
TRISC = %11111001          ; set PORTC.2, the speaker
                           ; connection
                           ; to an output line
TRISD = %00000000          ; set all PORTD lines to output
TRISE = %00000000          ; set all PORTE lines to output
PORTE.2=0                  ; set LCD to write mode
X VAR WORD                 ; set up the variable
ADCON1=%00000111           ; Set the Analog-to-Digital control
                           ; register
                           ; needed for the 16F877A see notes
PAUSE  500                 ; pause for LCD to start up
LCDOUT $FE, 1              ; clear screen
                           ;
ON INTERRUPT GOTO INT_ROUTINE  ; tells program where to go on
                           ; interrupt
T1CON=%01110001            ; sets us the interrupt enable
PIE1.0=1                   ; clears the interrupt flag so it can
                           ; be set
INTCON.6=1                 ;
OPTION_REG=%00000111       ; sets the prescalar to 256
X=0                        ; sets the initial value for X
                           ;
MAIN:                      ; the main loop of the program
  LCDOUT $FE, $80, "Metronome"  ; display first line
  LCDOUT $FE, $C0, DEC5 X  ; write X to line 2
  IF X=15 THEN             ; check value of X
    X=0                    ;
    TOGGLE PORTC.2         ; toggle the speaker so we can hear
                           ; it
  ENDIF                    ;
GOTO MAIN                  ; repeat loop
                           ;
DISABLE                    ; req'd Instruction to the compiler
INT_ROUTINE:               ; interrupt service routine
  X=X+1                    ; increment the counter
  PIR1.0=0                 ; clear the interrupt flag
RESUME                     ; go back to where you were
ENABLE                     ; reqd Instruction to the compiler
                           ;
END                        ; all programs must end with End
```

Timer2

Timer2 is discussed on page 59 in the datasheet.

Basic description: Timer2 has 8 bits, and both a pre- and postscalar. It uses the internal clock only and shares the prescalar with the watchdog timer. It can be turned on and off but cannot be used as a counter. It cannot be used as a counter because it does not have a pin on which it can receive the counter signals.

Timer2 is an 8-bit timer with both a prescalar and a postscalar. The prescalar can be as great as 16, and the postscalar can be as great as 16, for a total scaling of $16 \times 16 = 256$, which is the same as for Timer0. However, not all 256 scalings are available as they are for Timer0 (because the prescalar can only be set to 1, 4, or 16).

The Timer2 is identified as register TMR2.

The control registers for Timer2 are INTCON, T2CON, PIE1, and PIR1. The bits in these registers control Timer2 as follows:

T2CON.1 and T2CON.0 set the value of the prescalar as follows:

Bit value	Prescalar value selected
00	1, No prescalar is used
01	4
1x	16

The postscalar is specified by the 4 bits from T2CON.6 to T2CON.3 as follows:

Bit value	Postscalar value selected
0000	1, No prescalar is used
0001	2
0010	3
0011	4
0100	5
0101	6
0110	7
0111	8
1000	9
1001	10
1010	11
1011	12

1100	13
1101	14
1110	15
1111	16

~~(The T2CON.7 bit is not used and is read as a 0)~~
PIE1.1 and INTCON.6 control the interrupt enable bit
The interrupt flag is located at PIR1.1.

Like all the timers, Timer2 is clocked by the crystal frequency divided by 4 (or Fosc/4). It can *not* be used with an external oscillator.

The timer is controlled by T2CON.2 and is on when this bit is set to 1.

The Timer2 Program

The listing in Program 16.6 shows how Timer2 can be used to create a metronome. Compare this to the listings for Timer0 and Timer1 mentioned previously.

Program 16.6   Metronome using Timer2

```
; Like we did for TIMER0, first let us set up the LCD
CLEAR                         ; always start with clear
DEFINE OSC 4                  ; define oscillator speed
DEFINE LCD_DREG PORTD         ; define LCD connections
DEFINE LCD_DBIT 4             ; 4 bit path
DEFINE LCD_RSREG PORTE        ; select reg
DEFINE LCD_RSBIT 0            ; select bit
DEFINE LCD_EREG PORTE         ; enable register
DEFINE LCD_EBIT 1             ; enable bit
LOW PORTE.2                   ; make low for write only
; Set the port directions. We must set all of PORTD and all of
; PORTE as outputs
; even though PORTE has only 3 lines. The other 5 lines will
; be ignored by the system.
; PORTC is needed because the piezo speaker is on pin 2 of this
; port.
                              ;
TRISC = %11111011             ; set PORTC.2 the speaker connection
                              ; to an output line
TRISD = %00000000             ; set all PORTD lines to output
TRISE = %00000000             ; set all PORTE lines to output
PORTE.2=0                     ; set LCD to write mode
```

(Continued)

Program 16.6 Metronome using Timer2 (*Continued*)

```
X VAR WORD                       ; set up the variable
ADCON1=%00000111                 ; set the Analog-to-Digital control
                                 ; register
                                 ; needed for the 16F877A see notes
PAUSE 500                        ; pause for LCD to start up
LCDOUT $FE, 1                    ; clear screen
                                 ;
ON INTERRUPT GOTO INT_ROUTINE    ; tells program where to go on
                                 ; interrupt
T2CON.5=%01111011                ; sets us the interrupt enable
; bit 0=1
; bit 1=1
; bit 2=0
; bit 3=1
; bit 4=1  Prescaler
; bit 5=1  Prescaler
; bit 6=1
; bit 7=0
PIE1.1=0                         ; interrupt enabled
INTCON.6=1                       ; permits interrupts to occur
X=0                              ; sets the initial value for X
                                 ;
MAIN:                            ; the main loop of the program
  LCDOUT $FE, $80, "Metronome"   ; display first line
  LCDOUT $FE, $C0, DEC5 X        ; write X to line 2 of the LCD
  IF X=15 THEN                   ; check value of X
    X=0                          ;
    TOGGLE PORTC.2               ; toggle the speaker so we can
                                 ; hear it
  ENDIF                          ;
GOTO MAIN                        ; repeat loop
                                 ;
DISABLE                          ; reqd instruction to the compiler
INT_ROUTINE:                     ; interrupt service routine
  X=X+1                          ; increment the counter
  PIR1.1=0                       ; clear the interrupt flag
RESUME                           ; go back to where you were
ENABLE                           ; reqd instruction to the compiler
                                 ;
END                              ; all programs must end with end
```

The Watchdog Timer

A fourth timer in the 16F877A is used as a watchdog timer. The function of a watchdog timer is to provide an interrupt if a program hangs up and thereby fails to periodically reset the watchdog timer. The watchdog timer interrupt flag is reset from time to time (always before it runs out) in the program to keep this from happening. The code for resetting the timer is provided by the compiler automatically. Check the appropriate option in the programming options.

The watchdog timer shares its prescalar with Timer0 on an exclusive basis, meaning that either the watchdog timer or Timer0 can use the prescalar (but not both). When you select the watchdog timer, the prescalar for Timer0 goes to 1:1. In fact, you select 1:1 for Timer0 by setting bit 3 in the option register, which at the same time assigns the prescalar to the watchdog timer.

17

UNDERSTANDING THE COUNTERS:
COUNTING MARBLES

Project 3

As you may have guessed, this is not really about counting marbles. It is about counting pulses that may represent any number of things you are interested in counting. These pulses may come slow or fast, regularly or randomly, or in some other way, and we need to know how to handle them in all situations. We use the counters for counting the events. The marbles just happen to be an interesting medium for us to play with: They are easily obtained, are easy to handle, and are inexpensive (and they do roll!). See Figure 17.1.

You need to know a few things about this problem before we start, however. They are

- This is not trivial by a long shot!
- There is a limit to how fast the marbles can go by in the real world, and the programs we create have to be faster than that.
- There is a considerable difference between counting marbles that go by one at a time in an orderly fashion and those that go by one next to the other, helter skelter, fast. In this project, we will count them one by one, and then in the later part of the project we will count them going by much faster.
- Signal bounce or signal jitter is a problem in any "real-world" situation. Especially so with these shiny glass marbles.
- As can be expected, there are many ways to get a job like this one done—meaning that not everything will have been explored when we are done with this chapter.

We have asked for and been given the following assurances from Mr. Marbles the owner and manager of the local Marbles Inc. plant. These assurances are designed to make the specifications of the project more complete and to make our task easier. (In other words, we have been given the parameters for what we are to do.)

234 UNDERSTANDING THE COUNTERS: COUNTING MARBLES

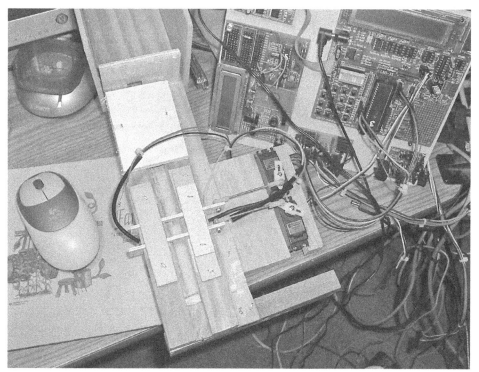

Figure 17.1 **Counting marbles, an escapement.** (Note the sensor between the two gates.)

- All the marbles are perfectly round.
- All marbles are the same size.
- All marbles have the same density.
- All marbles are opaque to infrared (IR) and light frequencies. (Here we are making the assumption that the IR light does not bounce around and cause us problems. This is not really a valid assumption!)
- All marbles are uniformly colored.
- We do not have to count more than 1000 marbles at a time.
- The maximum rate of marble flow is 900 marbles a minute (15 per second).
- The marbles will be given to us loose in a cardboard box and are to be returned the same way.

First, let's design a system that counts marbles by allowing them to go one by one only. In order for the marbles to be presented to the counter one by one, we need an escapement that will release them one at a time. The following is a schematic for a simple two-gate/lever escapement that we can make. It uses two model aircraft R/C servos to control the two gates that will let the marbles go by one at a time. This schematic is followed by photographs of a simple escapement that I made in my wood shop in less than an hour. A similar escapement could also be made out of cardboard or hobby materials glued together with hot glue if more comprehensive shop facilities are not readily available.

It is very important that you actually make the physical components needed for this project. They provide an important part of the "learning by doing" experience because they expose you to all sorts of things no one ever expects. It is what the real world is all about and it can all be done quite easily with some pieces of cardboard, a mat knife, and a hot glue gun. The idea is to build it. It does not have to be perfect. What you need to make is shown in Figure 17.2.

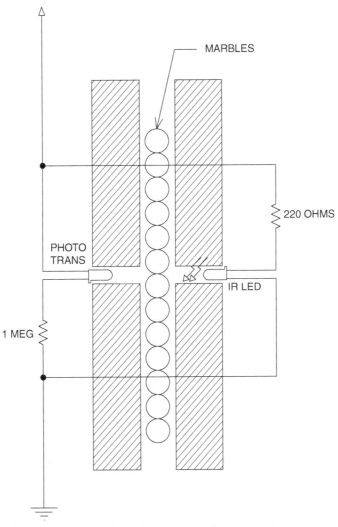

Figure 17.2 Schematic and photograph of a simple escapement. (Single sensor escapement for marbles using two servos. Sensor not shown. See Figure 17.1 for sensor.)

Figure 17.2 (*Continued*)

This simple escapement/marble chute will be used for all the counting projects with simple modifications and changes in sensor positions. Provisions should be made for a bin above and below the escapement to accommodate about a pint of marbles (around 100). The design should allow the marbles to be returned to the top bin between runs with ease so that experimental counting can proceed run after run without delay.

For the first escapement-based counting program, the escapement will be programmed to work as follows:

1. At startup, the lower gate is closed and the upper gate is open. The system ensures this by starting in this position. On reset, go to this position.
2. The counter is set to 0.
3. Load the marbles.
4. If there are any marbles in the system, the first marble comes to rest against the lower gate.
5. The *optical detector confirms this*.
6. If there is a marble above the lower gate, the upper gate closes.
7. Both gates are closed now.
8. There is a marble between the gates.
9. The lower gate opens and releases the marble.
10. The counter increments by one *at this point*.
11. The lower gate waits 0.01 seconds and closes again. This makes sure the marble has dropped away.

12. As soon as the lower gate has closed, the upper gate opens and stays open till a marble comes to rest against the lower gate. The *optical detector will confirm this*.
13. The cycle repeats as long as more marbles come down the chute.

This is the basic escapement model for almost all escapements. See Figure 17.3 for a simple implementation.

Notice that the system is designed in a way that does not leave any marbles behind in the counter at the end of the operations. This is an important consideration.

The system could be designed with only one servo, but we will use two servos, one for each gate so we can have complete and flexible control of the action. (Mind experiment: Think about how you would do this with only one servo.)

As mentioned earlier, we need a sensor to tell us if there is a marble resting on the lower gate. We will use an inexpensive IR transmitter/detector pair across the marble to provide this function.

We also need sensors to tell us when the two gates are fully open and fully closed, but since we are using servos, we can use the commands to the servos combined with suitable delays to know what the status of the gates is at any time. (We are assuming that the servos are always able to follow the commands given to them. If we were designing a much faster and more industrial pneumatic system, we might use micro switches, light beams, or proximity switches to confirm the in and out positions of the gates.)

Figure 17.3 Close-up view showing how the simple gates are made.

The connections to the system will be as follows:

Servo 1	J7 on the LAB-X1
Servo 2	J8 on the LAB-X1
Phototransistor	PORTB.0 on the LAB-X1
Infrared LED	PORTD.1 on the LAB-X1

SOME SPECIAL REQUIREMENTS FOR SERVOS

We need a way to pulse the servos on a continuing and regular basis. We know from the servo literature that the servos have to be reminded of their position about 60 times a second in order to maintain proper operation. We will need to set up an interrupt to come ON about 60 times a second to do this. The interrupt routine refreshes the counters in the servo every time it is called. The POT0, POT1 numbers in the routine for each servo represent the IN and OUT position of each of the servos.

FINDING THE SERVO POSITIONS

Note *Certain aspects of working with model aircraft servos concern the length and frequency of the signals they need. If you are not familiar with these, go to the Internet and read up on the relevant part on aircraft servomotor control.*

Before we can go any further, we need to determine the exact IN and OUT position of each gate so we can set these parameters to the appropriate values for each servo in the program.

Program 17.1 lets you move *both* servos under computer control from POT0 and watch the setting on the LCD. With this program, we can find the values of TOP_GATE and BOT_GATE for each of the servos with ease whenever we need to. The program uses the two lower potentiometers on the LAB-X1 to control the position of the servos in real time. Adjust the potentiometers to get the useable IN and OUT position of the servos and write the values down for each servo.

Program 17.1 This is a stand-alone program for finding the exact servo settings to determine the open and close positions of the gates

```
CLEAR                           ; clear the variables
DEFINE OSC 4                    ; define the oscillator
DEFINE LCD_DREG PORTD           ; define LCD connections
DEFINE LCD_DBIT 4               ; 4 bit protocol
DEFINE LCD_RSREG PORTE          ; register select byte
DEFINE LCD_RSBIT 0              ; register select it
DEFINE LCD_EREG PORTE           ; enable port
DEFINE LCD_EBIT 1               ; enable bit
```

(Continued)

Program 17.1 This is a stand-alone program for finding the exact servo settings to determine the open and close positions of the gates *(Continued)*

```
LOW PORTE.2                       ; leave low for write
ADCON1=%00000111                  ; set A-to-D control register
PAUSE 500                         ; pause for .5 sec for LCD
                                  ; startup
                                  ;
TRISC = %00000000                 ; PORTC is all outputs for servos
PORTC = %00000000                 ; set to 0
PORTD = %00000000                 , PORTD is all outputs for LCD
PORTE = %00000000                 , PORTE is all outputs for LCD
LCDOUT $FE, 1, "CLEAR"            ; clear display and show CLEAR
DEFINE ADC_BITS 8                 ; set number of bits in result
DEFINE ADC_CLOCK 3                ; set internal clock source
                                  ; (3=rc)
DEFINE ADC_SAMPLEUS 50            ; set sampling time in us
POT0 VAR BYTE                     ; create adval to store result
POT1 VAR BYTE                     ; create adval to store result
                                  ;
LOOP:                             ; main loop
  LCDOUT $FE, $80, DEC3 POT0," ", DEC3 POT1," "   ; display data
  ADCIN 0, POT0                   ; read PORTA.0
  ADCIN 1, POT1                   ; read port a1
  POT0=POT0 + 23                  ; so actual pulse is displayed
                                  ; (arbitrary value)
  PULSOUT PORTC.1, POT0           ; pulse port c.1
  PAUSE POT1                      ; pause 1/60 seconds (approx
                                  ; value is 24)
  PULSOUT PORTC.0, POT0           ; pulse port c.2
  PAUSE POT1                      ; pause 1/60 seconds, (approx
                                  ; value is 24)
GOTO LOOP                         ; do it again
END                               ; we always end with end
```

Noted for the record: in FUTABA systems, the servo center is defined as a 1.52 milliseconds wide pulse delivered about 60 times a second. The positional range is about 0.75 milliseconds on either side of that. Other manufacturers specify other values around 1.5 milliseconds, so it is worth it to check exactly what your servos need. Develop the habit of working to accurate parameters and looking things up for yourself.

In the preceding program, the pulses to both connectors *J7 and J8 use the same value, POT0.* Doing it this way allows us to hook a servo up to either connector, and the operation will be the same. Or both servos can be connected at the same time and you can find the end positions for both with this program. Both servos will not have the same value for the open and close positions for their respective gates unless you can make the linkages absolutely identical. *POT1 controls the delay* between the pulses. We want to find the shortest and longest delays that give us smooth servo operation, and then use the average.

Counting with an Escapement

Let's develop the "escapement-based" counting program step by step, segment by segment, as we were just discussing. At the start of the program, we need to wait for a signal to start the counting process. At this point, the top gate must be open and the lower gate must be closed. When the first marble arrives on top of the bottom gate, we will be ready to start counting. We assume the other marbles are in position, but this—as we will see—may not necessarily be the case, meaning that our software must accommodate all possibilities.

The hardware we will respond to and control consists of the following items connected as indicated:

Infrared LED power on PORTD.1

Infrared LED ground to ground

Phototransistor power on PORTB.1

Phototransistor ground to ground

Phototransistor signal to PORTB.0

Top servo on PORTC.1 to the servo pins

Bottom servo on PORTC.2 to the servo pins

We will use switch 1 connected between line PORTB.4 and ground to start the counting process. PORTB.4 will be pulled up internally with OPTION_REG.7=0

The wiring schematic for the preceding description is shown in Figure 17.4.

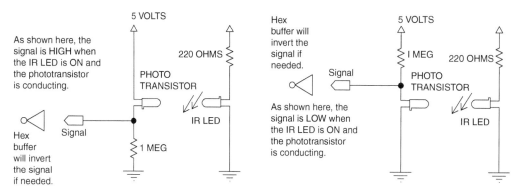

Figure 17.4 The two ways of doing schematic wiring diagrams for IR-LED phototransistor pairs.

Program 17.2 Program to count marbles with an escapement

```
CLEAR                           ; always start with clear variables
DEFINE OSC 4                    ; always define osc speed
; As always we will define the code for the LCD operation first.
DEFINE LCD_DREG PORTD           ; define LCD connections
DEFINE LCD_DBIT 4               ; and bits
DEFINE LCD_RSREG PORTE          ; select register
DEFINE LCD_RSBIT 0              ; select bit
DEFINE LCD_EREG PORTE           ; enable register
DEFINE LCD_EBIT 1               ; enable bit
LOW PORTE.2                     ; low mean we will write only
DEFINE LCD_COMMANDUS 2000       ; delay in micro seconds
DEFINE LCD_DATAUS 50            ; delay in micro seconds
DEFINE ADC_BITS 8               ; set number of bits in result
DEFINE ADC_CLOCK 3              ; set internal clock source (3=rc)
DEFINE ADC_SAMPLEUS 50          ; set sampling time in us
                                ;
; Next we set all the ports to the required directions
; and set their bits high and low
                                ;
TRISB=%11110000                 ; set PORTB direction
PORTB=%00001110                 ; set PORTB bits
TRISC=%00000000                 ; set PORTC direction
PORTC =%00000000                ; set PORTC bits
TRISD=%00000000                 ; set PORTD direction
PORTD =%00000000                ; set PORTD bits
PORTE =%00000000                ; set PORTE direction
; Then we set the registers to make the system digital
; Set the interrupt controls for the timer
; And the options register for the pullups and the prescalar
                                ;
ADCON1 =%00000111               ; set A, E registers to digital
INTCON=%10100000                ; set interrupt control reg bits
OPTION_REG=%00000101            ; set option reg bits
                                ;
; We need for the program to pause for half a second
; To allow the LCD to come up to speed and clear the screen
                                ;
PAUSE 500                       ; pause for LCD startup
LCDOUT $FE, $01                 ; clear the LCD
                                ;
```
Assign variables for the servos, and counters and set their values
```
                                ;
POT0 VAR BYTE                   ; create adval to store result
POT1 VAR BYTE                   ; create adval to store result
POT2 VAR BYTE                   ; create adval to store result
ALPHA VAR BYTE                  ; counter variable
```

(Continued)

Program 17.2   Program to count marbles with an escapement (*Continued*)

```
COUNTER VAR BYTE              ; counter variable
COUNTER=0                     ; set counter
                              ;
; Tell the program where to go on an interrupt
                              ;
ON INTERRUPT GOTO INT_ROUTINE ; set interrupt target
                              ;
; Set the initial conditions for the two gates as was determined
POT0=160                      ; close top gate
POT1=100                      ; close bottom gate
                              ;
; Create a loop that the program can wait in till we press
; Switch 1
                              ;
WAITLOOP:                     ; this is the loop for waiting to start
  IF PORTB.4=1 THEN           ; switch is connected to
    GOTO WAITLOOP             ; line B4
  ELSE                        ;
  ENDIF                       ;
  RETURN                      ;
; Since we will be reading the photo-transistor on port B0 we
; have to set PORTB
; so that B0 is an input.
; We are going to use the interrupt feature of B0 in later
; programs so we want to keep the sensor on B0.
; We could have use any line
; at this stage in this program
                              ;
TRISB =%11111111              ; we have to reassign this here for
                              ; B0 input
                              ;
LOOP:                         ; main loop
  PORTD.3=PORTB.0             ; so you can see the signal on D3
  POT0=120                    ; open top gate
  LCDOUT $FE, $80, DEC3 POT2," ","PRESS RESET"   ;
  LCDOUT $FE, $C0, DEC3 COUNTER,"     "          ;
  IF PORTB.0=0 THEN LOOP      ; if no marble then go back to loop
  ADCIN 0, POT2               ; read POT0 for servo wait pause
  POT0=158                    ; open top gate
  POT1=160                    ; close bottom gate
  GOSUB PAUSER                ; pause between moves
  POT0=125                    ; close top gate
  POT1=120                    ; open bottom gate
  GOSUB PAUSER                ; pause between moves
  COUNTER=COUNTER + 1         ; count the marble
GOTO LOOP                     ; go back to top
                              ;
```

(*Continued*)

Program 17.2 Program to count marbles with an escapement (*Continued*)

```
DISABLE                     ; disable interrupts
INT_ROUTINE:                ; the interrupt routine is called
                            ; every 1/60
  PULSOUT PORTC.1, POT0     ; of a second and pulses both servos
                            ; to
  PULSOUT PORTC.0, POT1     ; their set values. Then it
  INTCON.2=0                ; resets the interrupt flag for Timer0
RESUME                      ; resume program
ENABLE                      ; enable interrupts
                            ;
PAUSER:                     ; this routine is read from POT0 and
  FOR ALPHA =1 TO POT2      ; determines how long it is between
    PAUSE 2                 ; instructions to the servos.
  NEXT ALPHA                ; adjust it with POT0
RETURN                      ; do not make it to short
END                         ; All programs end in END.
```

The program starts out with both gates closed.

The code in Program 17.2 will stay in the "waitloop" until switch 1 is pressed.

That opens the top gate and the system is ready to count.

As soon as a marble stops against the lower gate, the marble is detected, the top gate closes and the bottom gate opens.

One is added to the counter (*at this point and not any sooner or later*).

The bottom gate closes.

The top gate opens and the cycle is repeated.

When there are no more marbles, the system stops.

Press reset to start over.

Some Real-World Notes

If we were designing for a real-world situation, we would add some type of detectors to make sure the gates actually went in and out as intended. Proximity detectors of some sort are the detectors of choice in that they have no moving parts. We would probably use pneumatic actuators and run the operation from a programmable logic controller (PLC), though as we saw in this exercise, a PIC would be much cheaper and would do the job just fine. The system could also, of course, be designed to use two solenoids with appropriate IN/OUT detectors.

The preceding is the simplest way to count the marbles, but it is *painfully slow*. Slow enough to be counted with an expensive PLC!

Counting to a Register Using an Interrupt

When the marbles are coming fast and furious, we cannot stop them one at a time and count them as we did in the previous program. We need a faster way of doing the job. One way is to have each marble cause an interrupt and then in the interrupt routine increment the variable that has the marble count in it. (Keeping in mind that the largest variable in PBP is 16 bits or 65,536, unless we make arrangements to count to a larger number by keeping track of the overflows at 65,536.)

The program keeps counting till the marbles run out. Reset takes you back up to the top. This process will be fast enough for most counting, but there are instances when it will not—a topic we will discuss later in this chapter.

The front end of Program 17.3 is very similar to the top of the preceding escapement program. The changes are in how we do the counting.

Program 17.3 This program is not executable (Program not for use in real counting as it is now; for discussion and examination only)

The top of the program now looks like this:

```
CLEAR                          ; always start with clear variables
DEFINE OSC 4                   ; always define osc speed
DEFINE LCD_DREG PORTD          ; define LCD connections
DEFINE LCD_DBIT 4              ; and bits
DEFINE LCD_RSREG PORTE         ;
DEFINE LCD_RSBIT 0             ;
DEFINE LCD_EREG PORTE          ;
DEFINE LCD_EBIT 1              ;
LOW PORTE.2                    ; low mean we will write only
DEFINE LCD_COMMANDUS 2000      ; delay in micro seconds
DEFINE LCD_DATAUS 50           ; delay in micro seconds
DEFINE ADC_BITS 8              ; set number of bits in result
DEFINE ADC_CLOCK 3             ; set internal clock source (3=rc)
DEFINE ADC_SAMPLEUS 50         ; set sampling time in us
                               ;
; Next we set all the ports to the required directions
; And set their bits high and low
                               ;
TRISB=%11110000                ; set port directions and values
PORTB=%00001110                ;
TRISC=%00000000                ;
PORTC =%00000000               ;
TRISD=%00000000                ;
PORTD=%00000001                ;
                               ;
; Then we set the registers to make the system digital
; Set the interrupt controls for the timer
```

(*Continued*)

Program 17.3 **This program is not executable** (Program not for use in real counting as it is now; for discussion and examination only) (*Continued*)

```
; And the options register for the pullups and the prescalar
                                ;
ADCON1=%00000111                ; set registers to digital
INTCON=%10100000                ; set interrupt control reg bits
OPTION_REG=%00000101            ; set option reg bits
; We need for the program to pause for half a second
; To allow the LCD to come up to speed and clear the screen
                                ;
PAUSE 500                       ; pause for LCD startup
LCDOUT $FE, $01                 ; clear the LCD
                                ;
; Assign variables for the servos and counters and set the values
                                ;
POT0 VAR BYTE                   ; create adval to store result
POT1 VAR BYTE                   ; create adval to store result
POT2 VAR BYTE                   ; create adval to store result
ALPHA VAR BYTE                  ; counter
                                ;
; Tell the program where to go on an interrupt
ON INTERRUPT GOTO INT_ROUTINE   ; set interrupt target
                                ;
; Set the initial conditions for the two gates as top open,
; bottom closed
POT0=125                        ; open top gate
POT1=100                        ; close bottom gate
                                ;
; Create a loop that the program can wait in till we press Switch 1
WAITLOOP:                       ; this is the loop for
PORTD.3=PORTB.4                 ; the start switch to be pressed
IF PORTB.4=1 THEN               ; switch is connected to
  GOTO WAITLOOP                 ; line B4
ELSE                            ;
ENDIF                           ;
                                ;
; Since we will be reading the photo-transistor on port B0 we
; have to reset
; TRISB for PORTB so that B0 is an input. We are going to use
; the interrupt feature of
; B0 in later programs so we want to keep the sensor on B0. We
; could have
; used any line at this stage in this program. We need to make
; B0 an input so
; that it can receive the signal from the IR - Phototransistor
; pair and enable
; the interrupt capability for B (later).
                                ;
```

(*Continued*)

Program 17.3 **This program is not executable** (Program not for use in real counting as it is now; for discussion and examination only) (*Continued*)

```
TRISB =%00001111           ; we have to reassign this here
                           ; for B0 input
OPTION_REG =%01111111      ; set bits
INTCON=%10010000           ; set bits
TMR1L =0                   ; clear low byte
TMR1H =0                   ; clear high byte
                           ;
LOOP:                      ; we do the counting work here in
                           ; this loop
                           ; this is an empty loop for now
GOTO LOOP                  ; end of loop
                           ;
INT_ROUTINE:               ; show contents of loop here
LCDOUT $FE, $80, DEC3 TMR1H, " ", DEC3 TMR1L," "    ;
RESUME                     ;
END                        ;
```

This is an acceptable way of counting marbles, but we can do the job even faster if we use one of the timers in the counter mode. We will consider both Timer0 and Timer1. (Remember: Timer2 cannot be used as a counter because it has no input pin.)

Counting Directly into an Internal Counter

Using *Timer0* (8 bit) to learn about counting to counters.

Timer0 is not suitable for counting the marbles, so we will write a short program to demonstrate its use as a counter. We will set up a situation that will let us count how many times we press button 1 on the LAB-X1 keypad, starting with 250 in the counter. Doing this lets us see the counter overflow after a few button presses. We will also monitor the contents of INTCON on the LCD to see what happens as the button is pressed again and again. You need to master techniques like these to look at what is happening in the PIC or other processors your using, whenever necessary.

Set up the counter for Timer0 as follows:

```
OPTION_REG=%01101000       ; No prescalar and external clocking
                           ; on TOCK1 (bit 5), rising edge (bit 4)
INTCON=%00100000           ; Enable interrupts with bit5
```

TOCK1, which is pin 6 of the PIC, will be used as an input for the counter (PORTA.4). Provide a jumper from D.1 to A.4 so the result of the key press, which appears on D.1, will be routed to the counter.

The following is what you should see on the LCD:

COUNTING DIRECTLY INTO AN INTERNAL COUNTER

Every time you press button 1 on the keypad, the counter in TMR0 will increment by 1. After it gets to 255, the next key press will show bit 2 on INTCON go to 1 for 0.25 seconds. Next the counter will reset to 250, and the bit at INTCON.2 will be cleared. INTCON.0 is the *"PORTB change"* interrupt flag. It stays at 1 because we keep pressing a key that affects PORTB.

The program that reflects that above is shown in Program 17.4.

Program 17.4 Demonstration of the use of TMR0 as a Counter

```
CLEAR                       ;
DEFINE OSC 4                ; osc frequency
DEFINE LCD_DREG PORTD       ; define the LCD connections
DEFINE LCD_DBIT 4           ;
DEFINE LCD_RSREG PORTE      ;
DEFINE LCD_RSBIT 0          ;
DEFINE LCD_EREG PORTE       ;
DEFINE LCD_EBIT 1           ;
LOW PORTE.2                 ;
                            ;
TRISA = %00010000           ; set PORTA
TRISB = %11110000           ; set PORTB
TRISD = %00000000           ; set PORTD
TRISE = %00000000           ; set PORTE
ADCON1= %00000111           ; don't forget to set ADCON1
OPTION_REG=%01101000        ; sets interrupt enable for TMR0
; bit0=0 not relevant because bit 3 is 1
; bit1=0 not relevant because bit 3 is 1
; bit2=0 not relevant because bit 3 is 1
; bit3=1 sets prescalar to 1:1 or no scaling
; bit4=0 sets high to low transition on count
; bit5=1 sets counter mode using A.4 as input
; bit6=0 sets interrupt edge select
; bit7=0 sets pull ups on PortB
PAUSE 500                   ; pause .500 second for LCD to start up
LCDOUT $FE, 1               ; clear LCD, go to first line
TMR0=250                    ; put 250 into counter
INTCON=%00100000            ; clear all of interrupt register
                            ; except bit 5
                            ;
MAIN:                       ; main loop
  LCDOUT $FE, $80, BIN8 TMR0, "=", DEC3 TMR0,"=TMR0"     ;
  LCDOUT $FE, $C0, BIN8 INTCON ,"=","INTCON"             ;
  IF INTCON.2=1 THEN        ; If the interrupt flag is set
    LCDOUT $FE, $80, BIN8 TMR0, "=", DEC3 TMR0  ; show changes to
    LCDOUT $FE, $C0, BIN8 INTCON ,"=","INTCON bits" ; the two
                                                    ; registers
    TMR0=250                ; reset count in TMR0
    PAUSE 750               ; pause so you can see what happens
```

(Continued)

Program 17.4 Demonstration of the use of TMR0 as a Counter (*Continued*)

```
      INTCON=0              ; reset INTCON register
  ELSE                      ;
  ENDIF                     ;
GOTO MAIN                   ; do it again
                            ;
END                         ; end program.
```

In Program 17.4, we press SW1 again and again and see what happens in the LCD. For each key press, you need to both release and hold the button down for a good part of a second to actually observe the key-down and key-up effect so that the 750 pause that allows us to look at the LCD can be discounted.

Using Timer1 in Counter Mode

We will use *Timer1* as the counter for the marbles because it is the only timer that can count to more than 255 without having to count the interrupts at every overflow of the counter and make the related calculations. It will count to the 1000 marbles designated in Mr. Marbles' specification.

Actually, the microprocessor is much faster than the marbles coming out the chute, but we will pretend that we need to really hurry in our counting process for some other process we might encounter in the future.

If the marbles are coming so fast we do not have time to update the registers, we have to set up an internal counter and feed the signals that the interrupt provides directly into that counter. The signals will have to be debounced before they can be connected to the internal counter because these counters are extremely fast and will pick up every little jitter as a countable event.

THE BASICS OF COUNTING DIRECTLY INTO A TIMER BEING USED AS A COUNTER

In order to use a timer as a counter, we must set up the timer as usual and then change the input to the timer from the internal clock to the external signal and set the appropriate parameters for the incoming signal (rising edge, and so on).

For Timer1, the 16-bit counter, this is done as follows:

Setting Timer1 up as a counter.

```
T1CON.4 and T1CON 5 = 0  ; for no pre-scale
PIE1.0=1                 ; to enable interrupt
INTCON.6=1               ; also needed to enable interrupt
T1CON.0=1                ; to start the counter. We have to
                         ; connect the
                         ; incoming signal to PORTC.0
T1CON.1=1                ; selects PORTC.0 for input
T1CON.3=3                ; also needed to select PORTC.0 for input
```

Note 1 *The interrupt flag at PIR1.0 will be set when the counter overflows (and would have to be cleared in the interrupt service routine—our count will never get that far, but we still need to be aware of this!). We are not interested in the counter overflow in this particular application because we will never count past 65635, but if we were to count past that, the code to reset the interrupt flag and keep track of the overflows would have to be added to the program at the appropriate location. It might even be necessary to close one of the gates (meaning: stop a very fast process) before we got to the overflow to stop everything while we got this done.*

Study the wiring diagram for the marble counter (Figure 17.5) so you fully understand how the hardware is wired on the controller.

In this program, each marble interrupts the IR beam, which feeds into B.0, the interrupt receiving pin. Any change in B.0 causes an interrupt and de-bounces the incoming

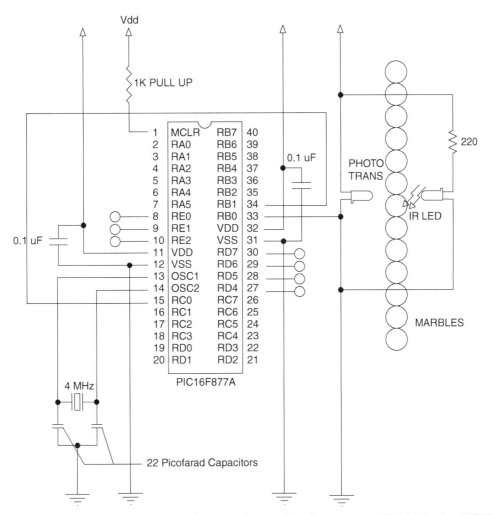

Figure 17.5 Wiring diagram for counting marbles into counter TMR1. (The four PORTD and the three PORTE lines, marked with circles in the figure, go to the LCD.)

signal because an interrupt can only be set once. The interrupt routine toggles pin B.4 up and down, and since pin B.4 is connected to pin C.0, the signal is fed to C.0. We are using pin C.0 because it is the only pin that can be used as an input for the Timer1 register pair as indicated in the data sheet.

THE WORKING PROGRAM

Program 17.5 demonstrates fundamental TMR1 operation as a counter for the marbles. As written it counts the number of times PORTB.0 is grounded. Displays show us the following things

```
;   Value of X
;   INTCON register  (too fast to see, add pauses if you want to
;   see it)
;   B register       (too fast to see, add pauses if you want
;   to see it)
;   TMR1L and TMR1H registers
;
; Working connections are
;   Connect B1 to C0 to feed Timer1 counter
;   Ground B0 to increment the counter.
```

The above ideas are implemented in Program 17.5.

Program 17.5 Basic marble counting routine to count directly to the Timer1 counter register pair (The wiring diagram for this program is in Figure 17.5)

```
CLEAR                          ; always start with clear. Good habit.
DEFINE OSC 4                   ; always define the osc speed. Good habit
DEFINE LCD_DREG PORTD          ; define LCD connections data PORTB
DEFINE LCD_DBIT 4              ; 4 bit path
DEFINE LCD_RSREG PORTE         ; register select port
DEFINE LCD_RSBIT 0             ; register select bit
DEFINE LCD_EREG PORTE          ; enable port
DEFINE LCD_EBIT 1              ; enable bit
LOW PORTE.2                    ; set low to write only
DEFINE LCD_COMMANDUS 2000      ; delay in micro seconds
DEFINE LCD_DATAUS   50         ; delay in micro seconds
ADCON1=%00000111               ; set ADCON1 for digital operation
                               ;
TRISA=%00000000                ; set PORTA
TRISB=%11110001                ; set port I/O
TRISC=%00001111                ; set PORTC
TRISE=%00000000                ; set PORTE
X VAR WORD                     ;
                               ;
T1CON=%00000011                ; set bits to control Timer1. Page 56
    ; bit7=0  Not used
    ; bit6=0  Not used
    ; bit5=0  Prescalar=1
    ; bit4=0  Prescalar=
    ; bit3=0  Osc enable bit. Osc off
    ; bit2=0  Synchronize to external clock
```

(Continued)

Program 17.5 Basic marble counting routine to count directly to the Timer1 counter register pair (The wiring diagram for this program is in Figure 17.5) *(Continued)*

```
             ; bit1=1  Use external clock
             ; bit0=1  Start Timer1. This bit turns in ON
OPTION_REG =%01111111      ; set bits. Page 21 of the datasheet
             ; bit7=0  Pulls up all of PORTB
             ; bit6=1  Interrupt edge select. Rising
             ; bit5=1  Clock source. RA4
             ; bit4=1  Increment on high to low
             ; bit3=1  Prescalar assigned to WDT therefore prescale=1
             ; bit2=1  Prescalar selection
             ; bit1=1  Prescalar selection
             ; bit0=1  Prescalar selection
INTCON     =%10010000      ; set bits for interrupt control.
                           ; Page 22
             ; bit7=1  Global interrupt enable
             ; bit6=0  enables any peripheral to set interrupt
             ; bit5=0  enables flag for TMR0 overflow. Goes with bit2
             ; bit4=1  Enables interrupt for bit 0 on PORTB. Goes with bit1
             ; bit3=0  enables interrupt any bit on PORTB. Goes with bit0
             ; bit2=0  TMR0 overflow flag or bit. TMR0 is always on.
             ; bit1=0  flag for bit0 on PORTB. See bit4
             ; bit0=0  flag for any bit on PORTB. See bit3
                           ;
PAUSE 500                  ; pause for start up
LCDOUT $FE, 1, "CLEARS THE LCD. ";  clear the display.
PAUSE 500                  ; this is for seeing a reset button
                           ; response
TMR1L=0                    ; clear low byte
TMR1H=0                    ; clear high byte
ON INTERRUPT GOTO INT_ROUTINE  ; interrupt target
                           ;
LOOP:                      ; main loop
  LCDOUT $FE, $80, "X=", DEC5 X,"       " ;
  LCDOUT $FE, $8A, "I=" ,BIN8 INTCON ;
  LCDOUT $FE, $C0, "B=",BIN8 PORTB," ",DEC3 TMR1H," ",DEC3 TMR1L
GOTO LOOP                  ; end of loop
                           ;
DISABLE                    ; disable interrupts
  INT_ROUTINE:             ; interrupt routine
    TOGGLE PORTB.1         ;
    TOGGLE PORTB.1         ;
    X=256*TMR1H + TMR1L    ;
    INTCON.1=0             ; reset the interrupt flag for B0
    PAUSE 25               ;
  RESUME                   ; resume program
ENABLE                     ; enable interrupts
                           ;
END                        ; end of program
```

UNDERSTANDING THE COUNTERS: COUNTING MARBLES

The count will be displayed as 2 three-digit single byte registers and will flow from the low byte on the right to the high byte on the left every time the count in the low byte spills over 255.

The counter increments every time line C.0 goes from 0 to 1. With the first marble resting on top of the lower gate, the count is 0. As soon as the marble drops away, the count goes to 1. It will stay at 1 till the second marble drops by. It will then go to 2 and so on.

The program needs to hold the lower gate closed and the upper gate open till you press SW1 to start the counting process. Once counting starts, the gates stay open till you reset the system at the end of the count.

The entire program with the "wait for SW1" added and the bit identification and some other comments removed is shown in Program 17.6.

Program 17.6 Program for counting to the TMR1 counter (See wiring diagram)

```
CLEAR                              ; always start with clear.
                                   ; Good habit.
DEFINE OSC 4                       ; always define the osc speed.
                                   ; Good habit
DEFINE LCD_DREG PORTD              ; define LCD connections data
                                   ; PORTB
DEFINE LCD_DBIT 4                  ; 4 bit path
DEFINE LCD_RSREG PORTE             ; register select port
DEFINE LCD_RSBIT 0                 ; register select bit
DEFINE LCD_EREG PORTE              ; enable port
DEFINE LCD_EBIT 1                  ; enable bit
LOW PORTE.2                        ; set low to write only
DEFINE LCD_COMMANDUS 2000          ; delay in micro seconds
DEFINE LCD_DATAUS   50             ; delay in micro seconds
ADCON1=%00000111                   ; set ADCON1 for digital
                                   ; operation
                                   ;
TRISB=%11110001                    ; Set port I/O
TRISC=%00001111                    ; set Port
X VAR WORD                         ;
                                   ;
T1CON=%00000011                    ; set bits to control Timer1
OPTION_REG =%01111111              ; set bits
INTCON=%10010000                   ; set bits for interrupt control.
                                   ;
PAUSE 500                          ; pause for startup
LCDOUT $FE, 1, "CLEARS THE LCD. "  ; clear the display.
PAUSE 250                          ; This is for seeing a reset
                                   ; button response
LCDOUT $FE, 1                      ; clear again.
TMR1L=0                            ; clear low byte
TMR1H=0                            ; clear high byte
ON INTERRUPT GOTO INT_ROUTINE      ; interrupt target
                                   ;
LOOP:                              ; main loop
```

(Continued)

USING TIMER1 IN COUNTER MODE 253

Program 17.6 Program for counting to the TMR1 counter (See wiring diagram)
(Continued)

```
    LCDOUT $FE, $80," TMR1= ",DEC3 TMR1H," ",DEC3 TMR1L    ;
    LCDOUT $FE, $C0, "COUNT=",DEC5 X    ;
GOTO LOOP                           ; end of loop
                                    ;
DISABLE                             ; disable interrupts
    INT_ROUTINE:                    ; interrupt routine
        TOGGLE PORTB.1              ;
        TOGGLE PORTB.1              ;
        X=256*TMR1H + TMR1L         ;
        INTCON.1=0                  ; reset the interrupt flag for B0
        PAUSE 20                    ;
    RESUME                          ; resume program
ENABLE                              ; enable interrupts
                                    ;
END                                 ; end of program
```

The wiring needed to implement the above program is shown in Figure 17.5.

RB0 will be high when the IR floods the phototransistor (which is the same as the dormant B.0 status, and this is what we want).

The wiring to test the operation of this program can be attached to the LAB-X1 and is shown in Figures 17.6 and 17.7.

The circuit can be mounted to the LAB-X1 and tested for proper operation before mounting to the marble counter. Play with a marble between the two devices to see how they react. The response to the edge of a marble is interesting, as is the point where you can induce the most jitter in the signal. See what happens when you change the value of the resistor in line with the phototransistor. High values are needed.

Adding the necessary startup sequence and switch 5 (SW5) to start the system, we get Program 17.7.

Program 17.7 Final program for counting marbles into the TMR1 counter
(Connect the servo to J7)

```
CLEAR                               ; always start with clear.
                                    ; Good habit.
DEFINE OSC 4                        ; always define the osc speed.
                                    ; Good habit
DEFINE LCD_DREG PORTD               ; define LCD connections data PORTB
DEFINE LCD_DBIT 4                   ; 4 bit path
DEFINE LCD_RSREG PORTE              ; register select port
DEFINE LCD_RSBIT 0                  ; register select bit
DEFINE LCD_EREG PORTE               ; enable port
DEFINE LCD_EBIT 1                   ; enable bit
LOW PORTE.2                         ; set low to write only
DEFINE   LCD_COMMANDUS 2000         ; delay in micro seconds
DEFINE   LCD_DATAUS   50            ; delay in micro seconds
```

(Continued)

Program 17.7 Final program for counting marbles into the TMR1 counter (Connect the servo to J7) (*Continued*)

```
DEFINE CCP1_REG PORTC        ;
DEFINE CCP1_BIT 2            ;
ADCON1=%00000111             ; set ADCON1 for digital operation
                             ;
TRISB =%11110001             ; set port I/O
PORTB=0                      ;
TRISC =%00001111             ; set Port
X VAR WORD                   ;
Y VAR BYTE                   ;
N VAR BYTE                   ;
M VAR BYTE                   ;
T1CON =%00000011             ; set bits to control Timer1
OPTION_REG  =%01111111       ; set bits.
INTCON=%10010000             ; set bits for interrupt control.
                             ;
PAUSE 500                    ; pause for startup
LCDOUT $FE, 1, "CLEARS THE LCD. "   ; clear the display.
PAUSE 250                    ; This is for seeing a reset
                             ; button response
LCDOUT $FE, 1                ; clear again.
TMR1L =0                     ; clear low byte
TMR1H =0                     ; clear high byte
ON INTERRUPT GOTO INT_ROUTINE   ; interrupt target
N=60                         ; Iteration count
M=16                         ; Pause between servo pulses
                             ;
LCDOUT $FE, $80, "HOLDING"   ;
FOR Y=1 TO N                 ;
    GOSUB GATE_IN            ;
NEXT Y                       ;
LCDOUT $FE, $80, "PRESS SW5 TO START "    ;
                             ;
WAITROUTINE:                 ;
IF PORTB.4=1 THEN WAITROUTINE    ;
N=35                         ;
LCDOUT $FE, $80, "OPENING GATE "    ;
PAUSE 500                    ;
FOR Y=1 TO N                 ;
    GOSUB GATE_OUT           ;
NEXT Y                       ;
                             ;
LOOP:                        ; main loop
  LCDOUT $FE, $80, "TMR1= ",DEC3 TMR1H," ",DEC3 TMR1L," "-
  LCDOUT $FE, $C0, "COUNT=",DEC5 X ;
GOTO LOOP                    ; end of loop
                             ;
```

(*Continued*)

Program 17.7 Final program for counting marbles into the TMR1 counter (Connect the servo to J7) (*Continued*)

```
DISABLE                      ; disable interrupts
  INT_ROUTINE:               ; interrupt routine
    TOGGLE PORTB.1           ;
    TOGGLE PORTB.1           ;
    X=256*TMR1H + TMR1L      ;
    INTCON.1=0               ; reset the interrupt flag for B0
  RESUME                     ; resume program
ENABLE                       ; enable interrupts
                             ;
GATE_IN:                     ;
  HIGH PORTC.1               ;
  PAUSEUS 2300               ; CW
  LOW PORTC.1                ;
  PAUSE M                    ;
RETURN                       ;
                             ;
GATE_OUT:                    ;
  HIGH PORTC.1               ;
  PAUSEUS 750                ; CCW
  LOW PORTC.1                ;
  PAUSE M                    ;
RETURN                       ;
END                          ; end of program
```

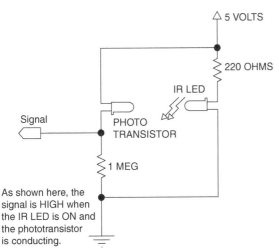

Figure 17.6 Diagram of the wiring needed to test the counter. (The high "normally open" condition here matches the weak high pullup on B.0 as it should.)

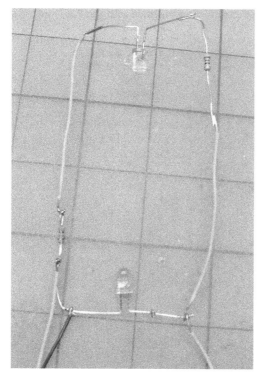

Figure 17.7 Wiring setup to test the sensitivity of the phototransistor response.

Special Notes for Timer1 Usage

Some of the succeeding paragraphs are from the datasheet.

TIMER1 OPERATION IN ASYNCHRONOUS COUNTER MODE

The following is what we use for our marble counting exercise.

If control bit T1SYNC (T1CON<2>) is set, the external clock input is not synchronized. The timer continues to increment asynchronous to the internal phase clocks. The timer will continue to run during sleep and can generate an interrupt on overflow, which will wake up the processor. However, special precautions in software are needed to read/write the timer (Chapter 6.4.1 of the datasheet). In asynchronous counter mode, Timer1 cannot be used as a time-base for capture or compare operations.

READING AND WRITING TIMER1 IN ASYNCHRONOUS COUNTER MODE

Reading TMR1H or TMR1L while the timer is running from an external asynchronous clock will guarantee a valid read (taken care of in hardware). However, the user should keep in mind that reading the 16-bit timer in two 8-bit values itself poses certain problems, *since the timer or one of the registers that form the timer may overflow between the reads.*

For writes, it is recommended that the user simply stop the timer and write the desired values. A write contention can occur by writing to the timer registers, while the register is incrementing. This can produce an unpredictable value in the timer register. Reading the 16-bit value requires some care. Examples 12-2 and 12-3 in the *PICmicro Mid-Range MCU Family Reference Manual (DS33023)* show how to read and write Timer1 when it is running in asynchronous mode.

THE TIMER1 OSCILLATOR

A crystal oscillator circuit is built in between pins T1OSI (input) and T1OSO (amplifier output). It is enabled by setting control bit T1OSCEN (T1CON<3>). The oscillator is a low-power oscillator rated up to 200 kHz. It will continue to run during sleeps. It is primarily intended for use with a 32 kHz crystal. The data sheet shows the capacitor selection for the Timer1 oscillator.

The Timer1 oscillator is identical to the LP oscillator. The user must provide a software time delay to ensure proper oscillator startup.

RESETTING TIMER1 USING A CCP TRIGGER OUTPUT

If the CCP1 or CCP2 module is configured in compare mode to generate a "special event trigger" (CCP1M3:CCP1M0 = 1011), this signal will reset Timer1. Timer1 must be configured for either timer or synchronized counter mode to take advantage of this feature. If Timer1 is running in asynchronous counter mode, this reset operation may not work. In the event that a write to Timer1 coincides with a special event trigger from CCP1 or CCP2, the write will take precedence. In this mode of operation, the CCPRxH:CCPRxL register pair effectively becomes the period register for Timer1.

RESETTING OF TIMER1 REGISTER PAIR (TMR1H, TMR1L)

TMR1H and TMR1L registers are not reset to 00h on a POR or any other reset except by the CCP1 and CCP2 special event triggers.

T1CON register is reset to 00h on a power-on reset or a brown-out reset, which shuts off the timer and leaves a 1:1 prescale. In all other resets, the register is unaffected.

It is recommended practice that you stop TMR1 before reading or writing the two registers. This avoids the problem of the low-byte register overflowing and changing the high-byte register while you are in the middle of setting the two registers. There is no way to set them both simultaneously. (This is possible in some other PICs.)

THE TIMER1 PRESCALAR

The prescalar counter is cleared on writes to the TMR1H or TMR1L registers so it has to be reselected after a write to the timer registers.

INTCON INTERACTION

If you watch INTCON on the LCD while interrupts are set and cleared, you will notice that the GIE (global interrupt enable) bit, bit7 of INTCON, gets cleared automatically when an interrupt is set, and is reset automatically when you clear the interrupt flag (if you had set it to 1 in the first place). This means that a second interrupt cannot occur before the first one that was set is cleared. Since PBP allows the use of more than one ON INTERRUPT GOTO command in a program, strategies to make sure all interrupts are handled properly can be formulated. It is best to do one interrupt at a time to keep it simple while you are a beginner.

TIMER2

The third timer is Timer2. This 8-bit timer *cannot be used as a counter* because it does not have the ability to respond to an external clock of any description. There is no line on this timer internal or external) that we can connect an external signal to.

18

A DUAL THERMOMETER INSTRUMENT

Project 4

Fabrication of the dual thermometer will be described in detail. Once the basic techniques used in making this instrument have been demonstrated, the projects that follow will be described in somewhat lesser detail. Later, we will convert this instrument into a controller, and then in the last project we will turn it into a data logger.

The dual thermometer device (see Figure 18.1) is based on the National Semiconductor LM34 temperature sensors. These inexpensive three-wire integrated circuits provide a voltage signal that can be converted to degrees Fahrenheit by multiplying that signal by 100. What could be simpler? What we must do is decide on the circuitry for hooking them up to the 16F877A PIC and then setting the appropriate registers. Like all other programs in this tutorial, the program will be written in PICBASIC PRO. The temperature readings are to be displayed on a 2-line-by-16-character display about every 0.2 seconds.

The datasheets for the LM34 and the 16F877A can be downloaded from the Internet.

The components shown in Table 18.1 are needed to make the dual thermometer.

See the circuit diagram for this instrument in Figure 18.2.

Note *PC boards, designed by me, that allow each of the relevant projects to be built on them are available from encodergeek.com. See the support Web site for pictures and other details. This boards fits in a box provided by All Electronics for a durable mounting for our projects. There is ample room in the box for potentiometers, switches, and so on to make other instruments on these boards*

The design of this instrument is an exercise in reading analog voltages, converting them to digital format, processing the information digitally, and displaying it on an LCD. As such, these are the basic steps required to read and display any analog signal with a PIC microprocessor, and are thus the basic processes for all our instruments.

There are eight pins on the 16F877A that can be used for analog input. They are distributed across ports A and E. Pin A.4 is excluded.

260 A DUAL THERMOMETER INSTRUMENT

Figure 18.1 Dual thermometer instrument.

Analog signal 0   PORTA pin 0

Analog signal 1   PORTA pin 1

Analog signal 2   PORTA pin 2

Analog signal 3   PORTA pin 3

                               PORTA pin 4   **Pin is not for analog use!** (see Table 18.2)

Analog signal 4   PORTA pin 5

Analog signal 5   PORTE pin 0

Analog signal 6   PORTE pin 1

Analog signal 7   PORTE pin 2

If you use the following ADCIN instruction:

`ADCIN 5, VAL_UE`

you will need to have the signal connected to pin E0 and the value read will be placed in variable VAL_UE. The number (5) in the instruction is a reference to the fifth analog signal number in these eight analog input pins, *not PIN A.5.*

Since most projects require a liquid crystal display, and we often use the lines on PORTE to control the display, the five lines on PORTA will be the ones most commonly used as analog inputs. (However, all three lines on PORTE can also be used as analog inputs if the LCD control lines are assigned to another port with the DEFINEs, or if you are not using the LCD at all.)

As with any PIC microprocessor with analog capabilities, the 16F877A microprocessor comes up in the analog mode on reset and startup. A number of the analog

TABLE 18.1 LIST OF ITEMS NEEDED TO MAKE THE DUAL THERMOMETER DEVICE

ITEM	QTY	DESCRIPTION
1	1	Experimenter's circuit board
		(available as a kit from encodergeek.com)
		Alternatives: RadioShack generic board; Velleman striped board
2	1	16-char-by-2-line display LCD
3	1	Microchip 16F877A microcontroller
4	1	20-MHz crystal
5	2	22 picofarad capacitors
6	1	2 feet of hookup wire
7	1	10 K resistor 1/8 watt for MCLR
8	1	7805 voltage regulator
9	1	1.0 uF cap 10 volts
10	1	0.1 uF cap 10 volts
11	2	LM34 temperature detectors
12	1	Battery holder with nuts and screws
13	1	Battery pigtail
14	1	Red LED
15	1	Switch
16	1	470 ohm 1/8 watt resistor for red LED
17	1	Bracket to mount switch
18	1	9-volt battery
19	X	Miscellaneous items from the junk drawer
20	2	Three pole screw connectors for sensors
21	1	40-pin socket for the PIC
22	1	2 × 5 pin programming connector
23	1	Box to hold everything

pins can be changed over to digital mode by setting the 4 lower bits in the ADCON1 register as indicated in Table 18.2.

The full table is on the next page.

Note that pin *PORTA.4 is not included* in Table 18.2. This pin is an open collector.

Pay special attention to how the reference voltages are specified at the various pins. See the datasheet.

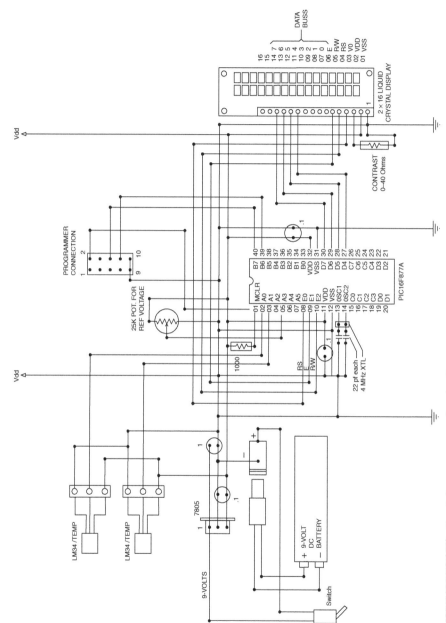

Figure 18.2 Dual thermometer circuit diagram.

TABLE 18.2 DIGITAL/ANALOG SELECTIONS MADE WITH THE ADCON1 REGISTER (/X/ IS USED TO INDICATE THE MISSING A.4 LINE)

PORT/PIN ANALOG	E.2 AN7	E.1 AN6	E.0 AN5	A.5 AN4	A.3 AN3	/X/	A.2 AN2	A.1 AN1	A.0 AN0	VREF+	VREF-	CH/REF
0000	A	A	A	A	/x/	A	A	A	A	Vdd	Vss	8/0
0001	A	A	A	A	/x/	Vref	A	A	A	RA3	Vss	7/1
0010	D	D	D	A	/x/	A	A	A	A	Vdd	Vss	5/0
0011	D	D	D	A	/x/	Vref	A	A	A	RA3	Vss	4/1
0100	D	D	D	A	/x/	A	D	A	A	Vdd	Vss	3/0
0101	D	D	D	D	/x/	Vref	D	A	A	RA3	Vss	2/1
0110	D	D	D	D	/x/	A	D	D	D	Vdd	Vss	0/0
0111	D	D	D	D	/x/	A	D	D	D	Vdd	Vss	0/0
1000	A	A	A	A	/x/	Vref	Vref	A	A	RA3	RA2	6/2
1001	D	D	A	A	/x/	A	A	A	A	Vdd	Vss	6/0
1010	D	D	A	A	/x/	Vref	A	A	A	RA3	Vss	5/1
1011	D	D	A	A	/x/	Vref	Vref	A	A	RA3	RA2	4/2
1100	D	D	D	A	/x/	Vref	Vref	A	A	RA3	RA2	3/2
1101	D	D	D	D	/x/	Vref	Vref	A	A	RA3	RA2	2/2
1110	D	D	D	D	/x/	D	D	D	A	Vdd	Vss	1/0
1111	D	D	D	D	/x/	Vref	Vref	D	A	RA3	RA2	1/2

NOTES

1. Settings 0110 and 0111 have identical results but are both shown, so all 16 combinations will be seen in the Table 18.2. The datasheet shows 011X for both lines.
2. Beware that AN7 to AN0 are the seven *analog input* identifications and Port E.2 to E.0 and A.5 to A.0 are *pin identifications* on the PIC.
3. *A.4 is missing* in this set, as stated earlier.
4. *Do not confuse the pins with the analog inputs*.

The preceding setting is described in detail on page 112 of the datasheet. Table 18.2 is a short form of that table. We do not have I/O access to lines A.7 and A.6 on the PIC 16F877A because they are internal to the processor. (They can, however, be read, and are related to the use of the parallel port capability of the PIC.)

In our particular case, we are interested in making pins A0 and A1 analog and using Vss and RA3 as the reference voltages to which we will compare the signal we receive from the LM34s. Since we have the LCD connected to PORTE, we need its pins to be digital. At this time, the status of the other lines is of no interest to us. The lines can be set to analog by selecting line 4 or 6 in the Table 18.2. For our purposes, which include using the LCD, all other lines can be digital. A useful selection is *line 4*, which makes all of PORTA analog and all of PORTE digital. In our case, PORTA.3 will be used as the reference voltage to which the incoming signals will be compared. We would specify that ADCON1 be set as follows:

ADCON1=%00000011 or ADCON1=3

(ADCON1=%00000111, which is used throughout this book to match the usual microEngineering Labs settings; it sets all the A and E lines to digital.)

As shown in the wiring diagram, the output from the two LM34s will be fed into the 16F877A at PORTA on lines A0 and A1.

Notice that the way we are wiring in the LM34s is identical to the way we wire in a potentiometer. In either case, the device is placed between the high- and low-power supply rails, and we read what is the equivalent of the wiper. This is the standard way of reading a voltage into a PIC microcontroller. If other than 0 to 5 volts are to be read in, appropriate voltage dividers and the necessary safety precautions have to be provided. In our case, we will use pin A.3 as the reference voltage, and the voltage on this pin can be adjusted with the potentiometer provided for this purpose. See the earlier diagram in Figure 18.2. The pins that the reference voltages are impressed on must be selected as indicated in Table 18.2—meaning that only A.2 and A.3 can be used.

Let's write the program to display the two signals on the 2-line-by-16-character display we are using for output. The display is to read as follows:

Temperatures
123°F   123°F

Provide two spaces on the left of the display so the display is centered in the LCD.

Let's start out by setting the DEFINEs to specify where the LCD is located. In the wiring shown in Figure 18.2, the data will be on the four high lines of PORTB, while the three control lines will be on PORTD. This arrangement is shown in Program 18.1.

Program 18.1   Code for a two-temperature instrument

```
CLEAR                       ; always clear variables
DEFINE OSC 4                ; always define the oscillator
DEFINE LCD_DREG PORTB       ; define LCD connections
DEFINE LCD_DBIT 8           ; 8 bit data path is faster
DEFINE LCD_RSREG PORTD      ; select register
DEFINE LCD_RSBIT 7          ; select bit
DEFINE LCD_EREG PORTD       ; enable register
DEFINE LCD_EBIT 6           ; enable bit
LOW PORTD.5                 ; we can leave this low for write only
                            ;
```

(Continued)

Program 18.1 Code for a two-temperature instrument (*Contitnued*)

```
DEFINE LCD_COMMANDUS 2000    ; delay in micro seconds
DEFINE LCD_DATAUS   50       ; delay in micro seconds
                             ; since we will be reading analog
                             ; lines,
                             ; we need the DEFINES to specify the
                             ; A-to-D conversions.
                             ; define the A2D settings
DEFINE ADC_BITS 8            ; set number of bits in result
DEFINE ADC_CLOCK 3           ; set internal clock source (3=rc)
DEFINE ADC_SAMPLEUS 50       ; set sampling time in uS
                             ;
                             ; set the Analog-to-Digital control
                             ; register
ADCON1=%00000011             ; needed for the 16F877A LCD
TEMP1 VAR BYTE               ; create adval to store first result
TEMP2 VAR BYTE               ; create adval to store second result
ADCON1=%00000011             ; set digital ports
PAUSE 500                    ; pause .500 second for the LCD
LCDOUT $FE, 1                ; clear the display
                             ;
LOOP:                        ;
ADCIN 0, TEMP1               ; read channel 0 to temp1
ADCIN 1, TEMP2               ; read channel 1 to temp2
TEMP1=TEMP1 * 100            ; convert to degrees F
TEMP2=TEMP2 * 100            ; convert to degrees F
LCDOUT $FE, $80, "Temp1=",DEC3 TEMP1," "  ; display first temp
LCDOUT $FE, $C0, "Temp2=",DEC3 TEMP2," "  ; display second temp
GOTO LOOP                    ; do it again
                             ;
END                          ; all programs must end with end
```

If the DEFINEs for the LCD are changed to the standard LAB-X1 display, Program 18.1 can be tested on the LAB-X1, and the positions of the wiper of POT0 and POT1 will be displayed on the screen. The necessary substitutions for the first few lines are shown in Program 18.2.

Program 18.2 Replacement code segment

```
CLEAR                        ; clear variables
DEFINE OSC 4                 ; osc speed
DEFINE LCD_DREG PORTD        ; data register
DEFINE LCD_RSREG PORTE       ; register select
DEFINE LCD_RSBIT 0           ; register select bit
DEFINE LCD_EREG PORTE        ; enable register
DEFINE LCD_EBIT 1            ; enable bit
DEFINE LCD_RWREG PORTE       ; read/write register
DEFINE LCD_RWBIT 2           ; read/write bit
DEFINE LCD_BITS 4            ; width of data
```

Using the same addresses used by the LAB-X1 lets us test our software on a system we know to be operating properly. Just knowing that the software will send something to the display can be heartening for a beginner. Once our confidence level regarding the use of these microcontrollers gets higher, we can abandon this strategy.

It is a good idea to take the time to add the ten-pin connector that allows us to program the PIC without removing it from the board we are working on. This makes it just about painless to make changes to the software, and in a really useful instrument (that you yourself have designed and built and thus know intimately about) this is a powerful feature that allows you to modify the characteristics of the instrument whenever you need to with a few keystrokes.

Another very useful adjunct to the hardware is to provide an additional input line and output line. These two lines can then be programmed to interact with the instrument we have created and would allow us to respond to a signal coming in on the input line and express the response on the output line after we had processed the information from the two sensors.

If we replace one of the sensors with a potentiometer, we can control the signal coming in on that sensor pin. Meaning, of course, that we can set one sensor input to whatever value we want with the potentiometer. This value can then be used as a set point that the other sensor reading is compared to before deciding whether we want to make our output line go high or low.

We have now turned our instrument into a very easy-to-set thermostat. If we like, this thermostat can be controlled from our auxiliary input line and the input line can get its signal from a remote source (maybe even over the Internet!). Keep in mind that we are not limited to using single line inputs and outputs. Our imagination is the only limit.

The instrument we created is more powerful and more flexible than that. The signal we send out can be a pulse width modulated (PWM) signal, where the level of modulation is a function of how far the set point is from the desired condition. It could be a frequency we might want to broadcast to the world—and so the beginnings of intelligent control start to come together.

The photographs in Figures 18.3 and 18.4 show the front and back of the prototype dual thermometer controller I built. Color versions of these photographs are much easier to read, and are on the support Web site that supports this book. On this board, the PIC is completely connected to the power side but none of the inputs/outputs are connected to anything except the LCD. This allows you to connect the I/O to whatever you want with jumpers to the screw terminals. Solder points are provided at the PIC end, as well as all the screw terminals, to make this easy.

In Figure 18.3, the universal project board has been mounted on a box, and a potentiometer and switch have been added to the system for future use.

Figure 18.4 shows the wiring needed to connect the I/O. We can also see the wiring to the potentiometer and the switch. These additions have been made in preparation for the next use of the instrument: data logging.

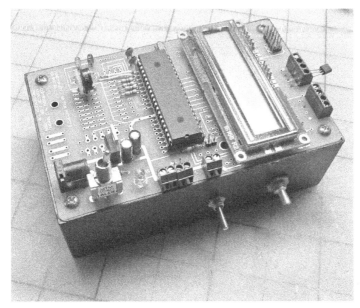

Figure 18.3 Dual thermometer instrument—front. (Either a one- or two-line display can be used.)

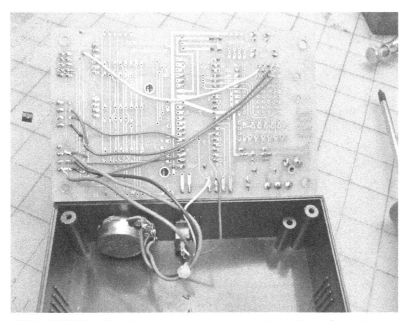

Figure 18.4 Dual thermometer instrument—bottom of board.

19

AN ARTIFICIAL HORIZON: A TABLE SURFACE THAT STAYS LEVEL

Project 5

In this project, we will read the inputs from a unique, inexpensive, two-axis gravity sensor and use them to control the horizontal position of a table.

We will use this very interesting and tiny electronic gravity sensor to build a table that stays horizontal while the base of the table is disturbed (within about 20 degrees) relative to the horizon. The sensor we will use is the Memsic 2125 dual-axis accelerometer. This sensor is available from Parallax Inc. already mounted on a tiny board with 6 pins that are 0.1 inches on centers.

The following is Parallax Inc.'s information on this Memsic accelerometer:

"The Memsic 2125 is a low cost, dual-axis thermal accelerometer capable of measuring dynamic acceleration (vibration) and static acceleration (gravity) with a range of ±2 g. For integration into existing applications, the Memsic 2125 is electrically compatible with other popular accelerometers."

Key features of the Memsic 2125 are.

Measures *0 to ± 2 g* on either axis; with a surprising resolution of less than *1 mg*. (This means that if you weigh 200 lb, this sensor could detect a change of 0.5 lb in your weight as affected by gravity.)

Fully temperature compensated over a 0°C to 70°C range

Simple pulse output of g-force for x and y axis

Analog output of temperature (T-Out pin)

Low current operation: less than 4 mA at 5 volts dc

A sampling of possible BASIC Stamp module applications with the Memsic 2125 include:
Dual-axis tilt sensing for autonomous robotics applications
Single-axis rotational position sensing
Movement/lack-of-movement sensing for alarm systems
R/C hobby projects such as autopilots

Memsic (www.memsic.com) provides the 2125 in a surface-mount format. Parallax mounts the circuit on a PC board providing all I/O connections so it can easily be inserted on a breadboard or through-hole prototype area.

The actuators we will use for the horizontal position correction of the table will be two radio control (R/C) hobby servos.

Discussion

There are two ways we can mount the sensor to correct the tilt of the table. We can mount the sensor on the base that is being tilted or we can mount the sensor on the actual table we want to keep level. (A third way would be to correct for the tilt in one direction by sensing base tilt, and in the other direction by detecting table tilt. This is an interesting academic exercise that we will not undertake here [but you may want to play with it on your own].)

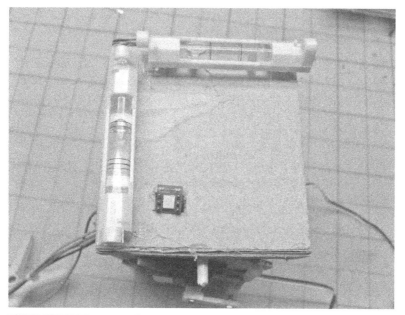

Figure 19.1 An artificial horizon–a basic unadorned cardboard horizontal table with a couple of levels and a gravity sensor. (Simple inexpensive construction is adequate for our purposes.)

If we mount the sensor on the base that will be tilting, we will get error signals that we will interpret, and then output the corrections to the target table. We may have to create a lookup table if there is any nonlinearity either in the response of the sensor or the mechanical linkages that connect the base to the top. In this case, the signal we detect is absolutely related to the position of the base. We are reading the actual error at the base. What we do with the signal is up to us, and how we design the linkages to level the table is also up to us. This is not the best way to do it in most cases, but this method may be the only one available to us in some situations.

If we mount the sensor to the top table, the signal we get will be a measure of how far the table has tilted. We can make the table come back to horizontal if we keep making an integrating correction till the table becomes horizontal; we stop when the error signal goes to zero. It is an integrative process over time. (If we wanted to know how large the correction was, we would have to keep track of how far we had moved the table to get it back to horizontal, but we do not need this information in an integrating system.)

Simply stated: it is better to mount the detector to the actual table because that is the surface we are interested in keeping level. It is best to get the error signal as directly as possible: at the source.

Setting Up the Hardware Connections

The instrument/controller we will create needs to read two inputs from the sensor and put out two outputs to the servos. The inputs are in the form of two frequencies we will read from the X-Out and Y-Out connectors of the sensor, and our outputs are the two 1 to 2 millisecond pulses we will output to the radio control hobby servos. See Figure 19.2.

The PBP language has a command (PULSOUT) that lets us output appropriate pulses on PORTC.0 and PORTC.1 on the LAB-X1, but in this exercise we will create our own pulses with the PAUSEUS command. However, we are still tied to PORTC.0 and PORTC.1 because that is how the circuitry for the LAB-X1 is laid out. The servo output pins on the LAB-X1 are hard wired to PORTC.0 and PORTC.1. The servos are fed from these two pins. If we were laying out a new board, we could choose to use any free port/pins.

We will use a ten-pin connector P2 on the LAB-X1 for all our connections because that is the connector that has PORTC.0 and PORTC.1 on it. This will allow us to use PORTC.2 and PORTC.3 as the inputs from the gravity sensor and so all the connections can be on one four-pin cable. The servos can plug into the standard servo connectors at J7 and J8. The circuitry for this is shown in Figure 19.2.

First, let's write a program to get a feel for what kind of pulses we are going to be getting from the Memsic sensor (we are interested in the pulse width, the frequency, and the range). To do this, we need to read the pulses from just one of the outputs and display them on the LCD as we move one axis though 180 degrees. Let's arbitrarily decide on the x axis output and write a program to read it. See Program 19.1.

272 AN ARTIFICIAL HORIZON: A TABLE SURFACE THAT STAYS LEVEL

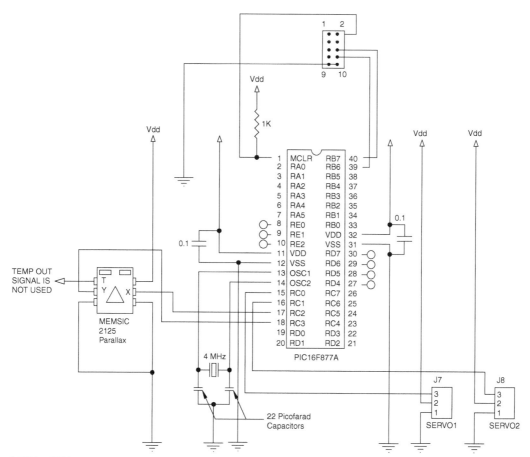

Figure 19.2 Wiring diagram for the stable table; connecting the Memsic accelerometer to two servos. (There are four wires between the sensor and the processor; both sides of Memsic must be grounded.)

What the literature tells us is that the output is a simple pulse output of the g-force. This is not very specific, so we will measure the length of the high portion of the pulse and the length of the low portion of the pulse. This will give us the total pulse cycle length (and the frequency, too). See Program 19.1

Program 19.1 Looking at the nature of the pulses received from the gravity sensor

```
CLEAR                        ; always start with clear
DEFINE OSC 4                 ; define oscillator speed
DEFINE LCD_DREG PORTD        ; define LCD connections
DEFINE LCD_DBIT 4            ; 4 bit path
DEFINE LCD_RSREG PORTE       ; select reg
DEFINE LCD_RSBIT 0           ; select bit
```

(Continued)

Program 19.1 Looking at the nature of the pulses received from the gravity sensor (*Continued*)

```
DEFINE LCD_EREG PORTE          ; enable register
DEFINE LCD_EBIT 1              ; enable bit
LOW PORTE.2                    ; make low for write only
ADCON1=%00000111               ; make ports digital
PAUSE 500                      ; pause for LCD startup
LCDOUT $FE, 1                  ; clear the LCD
TRISB=%11111111                ; set register for PORTC
X VAR WORD                     ; set variable x
Y VAR WORD                     ; set variable y
                               ; body of main loop
LOOP:                          ; start loop
  PULSIN PORTB.0, 1, X         ; measure pulse from Memsic
  PULSIN PORTB.0, 0, Y         ; measure pulse from Memsic
  LCDOUT $FE, $80, DEC5 X," ",DEC5 Y, " P Width"   ; print
                                                   ; conditions
  LCDOUT $FE, $C0, DEC5 X + Y, " Total"    ; print conditions
  PAUSE 50                     ; pause 1/20 seconds
GOTO LOOP                      ; back to loop
END                            ; always end with end
```

Surprise, surprise! The Memsic puts out a fixed frequency pulse with a variable duty cycle. Not exactly what you might expect from the information on the box the sensor comes in. On my sensor, I received a wavelength of almost exactly 1000 units (999 to 1001). Since I have the LAB-X1 set to 4 MHz, the resolution of the system is 10 μsec, and it gives a total cycle time of each cycle of 10,000 μsec, which is a frequency of 100 Hz. There is an important lesson in this: always check it out for yourself, because the instructions can be confusing.

The high side of the cycle varied from 380 to 620 units, or 500 + or − 120 units, as I moved the sensor around through 180 degrees.

The servos we are using require a center position pulse of 1520 μsec bracketed with a range of + or − 750 μsec. (You have to check this for your specific servos.)

The equation for converting what we read into what the servos need will be as follows:

Output pulse length = 1520 + (reading − 500)*5

We can implement the preceding conditions with the following for single-axis operation.

SINGLE-AXIS SOFTWARE

Let's first, as always, set up the DEFINEs for the LCD. We are using the LAB-X1 as our controller here so the standard port designations will need to be used. See Program 19.2.

Program 19.2 Program for single-axis artificial horizon

```
CLEAR                         ; always start with clear
DEFINE OSC 4                  ; define oscillator speed
DEFINE LCD_DREG PORTD         ; define LCD connections
DEFINE LCD_DBIT 4             ; 4 bit path
DEFINE LCD_RSREG PORTE        ; select reg
DEFINE LCD_RSBIT 0            ; select bit
DEFINE LCD_EREG PORTE         ; enable register
DEFINE LCD_EBIT 1             ; enable bit
LOW PORTE.2                   ; make low for write only
ADCON1=7                      ; make A, E ports digital
PAUSE 500                     ; pause for LCD startup
X VAR WORD                    ; set variable x
Y VAR WORD                    ; set variable y
TRISB=%11111111               ; set register for PORTC
TRISC=%11111100               ; set register for PORTC
LCDOUT $FE, 1                 ; clear LCD
                              ;
LOOP:                         ; start loop
PULSIN PORTB.0, 1, X          ; measure pulse from Memsic
PULSIN PORTB.1, 1, Y          ; measure pulse from Memsic
                              ;
LCDOUT $FE, $80, DEC4 X, " ", DEC4 (1520+6*(500-X))   ;
LCDOUT $FE, $C0, DEC4 Y, " ", DEC4 (1520+6*(500-Y))   ;
                              ;
PORTC.0 = 1                   ; start servo pulse
PAUSEUS   (1320+6*(500-X))    ; length of pulse for servo
PORTC.0 = 0                   ; end pulse
                              ;
PORTC.1 = 1                   ; start servo pulse
PAUSEUS   (1320+6*(500-Y))    ; length of pulse for servo
PORTC.1 = 0                   ; end pulse
GOTO LOOP                     ; back to loop
END                           ; always end with end
```

In Program 19.2, the multiplier, 6 in the PAUSEUS command, determines the sensitivity of the response. It is multiplying what is essentially the error signal. Try changing its value to 20 and see what happens. Try 10 and try 1. Selecting the right gain for the error signal is important.

The 500 is the at rest (horizontal) reading from the Memsic on my particular sensor. There may be slight variations in the value from sensor to sensor, so we may want to add a potentiometer to the circuit wiring. That would allow us to make an adjustment (say, to always bring this value to 500). On the servo I was using, the center position was 1320 μsec. The theoretical value for this is usually stated as 1520 μsec (by Futaba). Each servo can be expected to be slightly different, and the mounting position of the servo arm/horn to the servo also affects the mechanical response we will get. Here again, we could provide a trim potentiometer to adjust this value.

The preceding exercise demonstrates the relative ease with which we can make a fairly sophisticated instrument, like an artificial horizon, when we use a PIC microcontroller as our logic engine.

This instrument could easily be modified to show how may g-forces you went through as you turned a corner in a car. In this case, the sensing axis of the sensor would have to be placed left to right across the automobile, and a multiplier would have to be adjusted to give a reasonable display on the LCD.

TWO-AXIS SOFTWARE

To make a two-servo table that keeps the table top horizontal in both directions, we have to add the code for the second servo (see Program 19.3) The LCD part of the program does not change. After that...

Program 19.3 Program for the dual-axis artificial horizon

```
; Set up the variables to be used
ALPHA VAR WORD                      ; set variable alpha
BETA   VAR WORD                     ; second variable for second axis
                                    ; body of main loop
LOOP:                               ; start loop
PULSIN PORTC.2, 1, ALPHA            ; measure pulse from Memsic for x
PULSIN PORTC.2, 1, BETA             ; measure pulse from Memsic for x
                                    ;
LCDOUT $FE, $80, DEC4 ALPHA," ", DEC4 BETA, " Memsic"
; print conditions
LCDOUT $FE, $C0, DEC4 ((5*(ALPHA-493))+1310)," Pulse to servo"
; print conditions
                                    ;
PORTC.0 = 1                         ; start servo 1 pulse
PAUSEUS (1310+(5*(ALPHA-493)))      ; length of pulse for servo 2
PORTC.0 = 0                         ; end pulse 1
                                    ;
PORTC.1 = 1                         ; start servo 2 pulse
PAUSEUS (1310+(5*(BETA-493))) ; length of pulse for servo 2
PORTC.1 = 0                         ; end pulse 2
                                    ;
PAUSE 15                            ; pause about 1/60 seconds
GOTO LOOP                           ; back to loop
END                                 ; always end with end
```

Building the Artificial Horizon Table

The easiest way for us to build the table is to build two tables one above the other with the two controlled axes at right angles to one another. The problem is that the two axes will not be completely independent because of the mechanical interactions in a simple, and not very precise, cardboard construction. This can be taken care mathematically in the software with a lookup table or an equation depending on how well the mechanism responds to the tilt of the base. Since the correcting mechanism designed for each table setup will be different, the implementation of this will be left up to each individual constructor, but we will cover the general principals. If we put the sensors at the table surface, none of this matters because we are sensing the actual tilt of the table and this is beyond the effect of the linkages, and so on.

I think it is really important for you to actually build a table. It does not have to be anything fancy, and you will discover all kinds of things about controlling the table that just will not occur to you if you don't have a table to play with. Though the concept is losing currency among our educators, nothing beats working with your hands and your mind. They reinforce one another. I built everything in this book many times over!

Figure 19.3 shows a picture of a table I built out of cardboard. The table components were cut out from an old cardboard box with a box knife and a pair of scissors, and then glued together with hot glue. Since our interest is in the control of the table as opposed to the table itself, a simple cardboard model lets us do what we want at minimal cost and with minimal effort. Fortunately, the slight sloppiness and inaccuracies in the construction of the table lend themselves to software corrections. My table is shown in Figures 19.3 and 19.4.

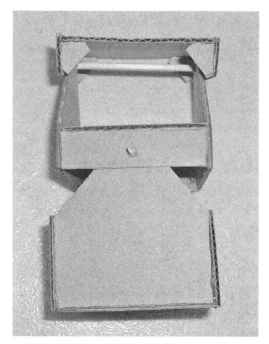

Figure 19.3 Picture of simple horizontal table and the gravity sensor; the basic table mechanism can be made out of cardboard.

Figure 19.4 Picture of the cardboard table mechanism with servos installed. (Mechanical inaccuracies will be compensated for by the software.)

The Memsic detector we are using needs four wires between the sensor and the PIC. A simple extension cable with a four-point female end can be made to fit to the high nibble of PORTC. Since we are using a relatively expensive sensor, we will not want to solder to it. Therefore, suitable slip-on connections should be provided on the sensor end also.

It is left up to you to modify the software to increase the speed with which the table returns to horizontal after the base has been moved.

Gravity Sensor Exercises

1. Make a sensor to investigate the gravity changes your body experiences as you travel up and down in a commercial elevator. How do the values vary between a hydraulic elevator in a small building and a traction elevator in a tall building. Were you surprised by what you found? (Investigate both the magnitude and duration of the accelerations.)
2. Investigate the gravity changes experienced by a passenger in a car. Both sideways motion and forward motion/acceleration should be studied.

20

BUILDING A SIMPLE EIGHT-BUTTON TOUCH PANEL

Project 6

TOUCH PANELS

Today, we see touch panels used everywhere. Every restaurant and store you visit has a handful of them. But how exactly do they work? Well, this chapter describes one way to skin this cat.

If you design an instrument or controller that needs a control panel in a hurry, by far the easiest way to make one is to do it with a touch panel. They are flexible and inexpensive and they go together fast. They are easily reprogrammed and most can be used with many applications just by changing the software. Depending on your requirements, the panel may be able to employ the PIC used by the instrument/controller you are designing, or it can be designed to have its own dedicated PIC.

In this project, we will build a touch panel to be placed over a plain piece of paper, with the functions written on the paper (see Figure 20.1). This panel will control the operation of two LEDs connected to the LAB-X1. We will control the operation of the LEDs from this panel. The following functions will be implemented. (Make a mental note that the control of a couple of small dc motors would be almost identical to this application.)

- Select the LED to address
- Turn LED on and off
- Select steady on or blinking for each LED
- Blink slower and blink faster for each LED

The principles involved in creating a touch interface for any number of applications are the same. Once you know how to create the interface on the LAB-X1, you will be able to add the interface to any surface. Though usually placed in front of a CRT, a touch interface is particularly well suited to being placed in front of printed and painted surfaces to

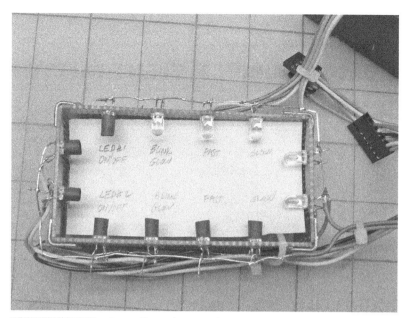

Figure 20.1 A small touch panel frame with IR emitters and photo-transistors in place.

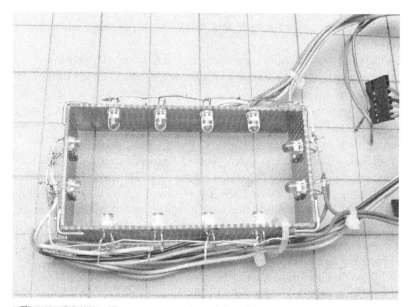

Figure 20.2 The components soldered into the touch panel frame.

replace any number of pushbuttons. It can be made relatively weatherproof easily and has no moving parts. Of course, the same principles also apply for touch panels placed in front of CRTs, and CRTs have the advantage that we can change what is displayed with each touch.

Our goal is to create a two-position-by-four-position interface for the LEDs. This will give us the equivalent of the eight buttons we need for the operations we have in mind. We will use the interface to control two LEDs that we will add to the LAB-X1. As mentioned earlier, we want to be able to turn each one on and off and be able to control the rate at which each of the LEDs blinks independently.

HARDWARE NEEDS

In order to create this eight-button interface, we need six infrared (IR) light beams, one for each of the two rows and one each for the four columns. Each IR light beam needs an emitter and a sensor. We will also need a simple frame to hold the 12 opto-electronic devices in place above the piece of paper describing the functions of the eight buttons. Figure 20.2 shows the scheme used by me in its initial phase with the components in place and ready to be wired together. Figure 20.3 shows the next phase in the wiring.

SOFTWARE NEEDS

The function of the eight buttons that the interface will simulate will have to provide the following operations for the two LEDs:

| LED 1 | ON/OFF | Steady or blink | Blink faster | Blink slower |
| LED 2 | ON/OFF | Steady or blink | Blink faster | Blink slower |

Simultaneously, we will display the conditions inside the LAB-X1 on the LAB-X1's LCD so we can actually see what is going on.

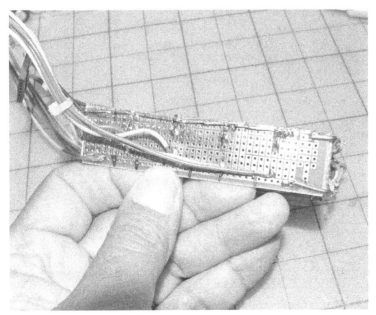

Figure 20.3 Detail wiring for the touch panel—close-up of one side. (For our immediate purposes, exposed wiring is acceptable.)

The operation of a touch interface is very similar to the scanning of a keyboard except that the wires and switches are replaced by infrared light beams and, of course, instead of making a contact with a switch, we break a light beam and detect the effect on a phototransistor. The problem is that light beams scatter and can illuminate more than one detector, while the wires go to one button only, making them easier to use. This means the scanning has to be more device-specific to be successful. We need to turn on one IR emitter at a time and look at the one phototransistor opposite it (ignoring all others). You will more clearly see what I mean by this as we go along.

Port usage: PORTA (two lines), PORTB (six lines), and PORTC (six lines) will be used for the project. Let's use PORTB (six lines) for connection to the six phototransistors, and use PORTC's (six lines) to power the six infrared LEDs one at a time. We will use the two lines (A.2 and A.3) on PORTA for the two LEDs. The wiring diagram for the touch screen as we will wire it is shown in Figure 20.4.

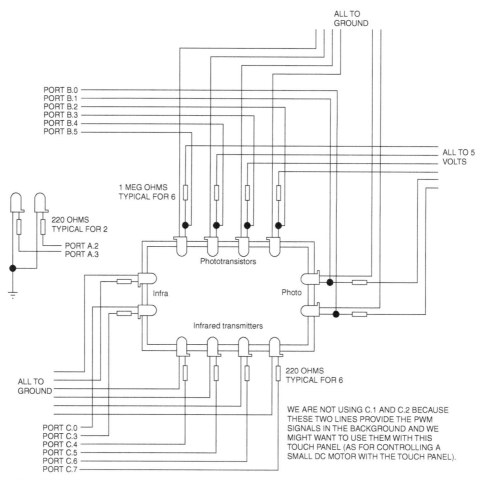

Figure 20.4 Touch panel schematic wiring diagram. (On the 16F877A, PORTD and PORTE are being used for the LCD.)

BUILDING THE TOUCH PANEL FRAME

Now that we know what we need in the way of hardware, let's build a simple touch frame. The easiest way to do this is to use a prefabricated PC board with lines on the back, and then cut out the parts you need to fit. Assemble the frame with the solder side out (as shown in Figure 20.5) and solder in the components. The IR beams that go across the LCD do not have to be at the same height as the ones that go across the short side but it will look better if they do.

The frame I made is approximately 4 inches long by 2 inches wide with sides 1 inch high. The infrared transmitters and phototransistors are to be placed on the half inch mark at every inch. I first assembled the frame and inserted the IR LEDs, the phototransistors and the resistors in place and make sure that they would all fit and still allow ample room for the wiring needed to connect everything up. Figure 20.4 shows how everything is wired together.

As shown in Figure 20.5, the LEDs and phototransistors are mounted in the printed circuit board segments that were cut from larger PC boards. The corners of the PC boards are held together with number 12 wire and then soldered firm. The wiring is extended from the LEDs and transistors back to the LAB-X1 ports as shown in the photos. Each wire has a push-on terminal attached to it that it is then placed in a header, which in turn fits on the pins that have been added to the LAB-X1 (see Figure 20.6). The two controlled LEDs can be mounted to the LAB-X1 board frame, or at a convenient location of your choice.

The basic scheme for using the IR transmitters with the photosensors is to turn them on and look at them a pair at a time in a suitable sequence and then to respond to the combination that indicates that a light beam that has been interrupted. The software

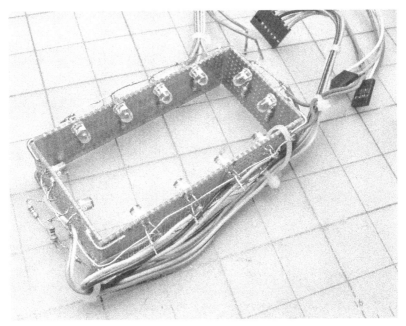

Figure 20.5 Touch panel ready for connection to LAB-X1.

Figure 20.6 Headers added to the LAB-X1 board for extension of pin signals to the panel.

pulls all of PORTB up with the internal pullups (OPTION_REG.7=0) and then grounds each line when the phototransistor conducts. Conduction takes place when the transistor is illuminated with IR. The circuitry for a one-line pair is shown in Figure 20.7.

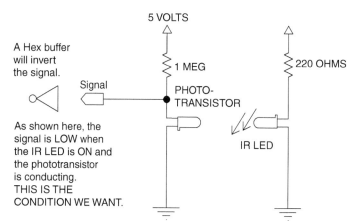

Figure 20.7 Single IR transmitter—phototransistor pair wiring. (Signal to the PORTB pin is on or high when beam is not broken.)

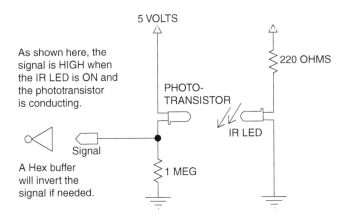

Figure 20.8 Single IR transmitter—phototransistor pair wiring.

Note *In the preceding case, we have pulled all the pins on PORTB high. So the dormant condition for our setup should keep the signal high when the light beam is not interrupted. When the phototransistor sees the IR, it conducts and its resistance goes to near zero and the signal goes to whatever side the phototransistor is connected to. In our case, this is 5 volts, so the signal is high when the IR is on and makes the transistor conduct. Figure 20.8 provides the opposite condition.*

The signal to the PORTB pin is off or low when the beam is not broken. *We do not want this condition.* The finished panel is shown in Figure 20.9.

Figure 20.9 The finished touch panel.

THE SOFTWARE

We will divide the software into three independent modules:

- Read which of the eight points has been selected with **getkey**.
- Display the information on the LAB-X1 so we can see what is going on with **display**.
- React to the selected key with **do_it**.

We need to set up the scan routine to determine which of the infrared beams have been interrupted so we can identify the area of the LCD that has been selected. The scheme for doing so is similar to the scheme used in scanning a keyboard. It consists of turning on the LED for column 1 and then checking to see if light beams for the column have been interrupted. We do this for each column one at a time. If a beam has been interrupted, the line associated with it will remain high. If it has not, we go on to the next line and check it. We make a record of the selection if a column was selected.

We do the same thing for the other row and then pass the information to the display routine, and then the execution routine. The execution routine uses the select case statement to choose the rows and column and react to them. After we are done, we reset the row and column variables to 0 and start over.

We start by creating all the code needed to activate the LAB-X1's liquid crystal display (see Program 20.1). This is the standard code we use in all our programs that use the LCD.

Program 20.1 Segment 1, LCD DEFINEs

```
CLEAR                       ; always start with clear
DEFINE OSC 4                ; define oscillator speed
DEFINE LCD_DREG PORTD       ; define LCD connections
DEFINE LCD_DBIT 4           ; data starts at bit 4
DEFINE LCD_DATA 4           ; 4 bit path
DEFINE LCD_RSREG PORTE      ; select reg
DEFINE LCD_RSBIT 0          ; select bit
DEFINE LCD_EREG PORTE       ; enable register
DEFINE LCD_EBIT 1           ; enable bit
LOW PORTE.2                 ; make low for write only
ADCON1=%00001110            ; set the ports for PORTs A, and
                            ; E to digital
PAUSE 500                   ; pause for LCD start up
LCDOUT $FE, 1               ; clear LCD
```

Next, let's set the data direction registers for all the ports we will be using. We will use PORTC for turning on the LEDs, and PORTB for detecting the condition of each of the phototransistors. PORTA will be used to turn the two LEDs on and off.

PORTD and PORTE are being used by the LCD, as indicated in Program 20.1.

Program 20.1 Segment 2—Set up the ports

```
TRISA = %00011001       ; Set PORTA
TRISB = %00111111       ; Set PORTB
TRISC = %00000110       ; Set PORTC
TRISD = %00000000       ; Set PORTD
TRISE = %00000000       ; Set PORTE
```

All the lines on PORTB are pulled high internally by the setting the option register as shown in Program 20.1

Program 20.1 Segment 3—Pull up PORTB

```
OPTION_REG.7=0
```

We are going to be looking at six of the eight lines on PORTB to determine if the touch panel has been accessed. (In our particular case, B.6,7 will not be used.) The six lines are connected as two rows of four columns to create the equivalent of an eight-button keypad. Check the wiring diagram again to see what is connected to what.

In this touchpad, we sense a selection when one of the column beams and one of the row beams is interrupted at the same time. The intersection identifies the selected area of the panel. In order to make sure one, and only one, area is selected, we turn on one column and then one row beam at a time, and if both of them are interrupted, we know we are at their intersection. We then turn on the next row on that column and check to see if both beams are interrupted and so on till all the rows and all the columns have been checked. We have to do it this way because the IR beams are diffuse and will illuminate more than one target when turned on. We eliminate all incorrect targets by looking at the intended target only.

In our wiring scheme, we turn the IR LEDs on in a sequence and look at the corresponding pin on PORTB to see if they have *not* been excited (= interrupted). On PORTC, the LEDs represented by 0 are turned on (the other side of the LEDs is tied to 5 volts). On PORTB, the pins will go 0 when on. (They were pulled up and showed as 1s when they were dormant.) We have to look at the six possible conditions and respond to each of them.

When the infrared illuminates the phototransistor, the transistor conducts and pulls the associated line on PORTB low. If your finger interrupts the beam, the PORTB line will become high. *That is what we are looking for.*

The rest of Program 20.1 follows, with extensive annotations:

Program 20.1 Final segment to make a finished program—the touch panel

```
X VAR WORD              ; counter variable
Y VAR BYTE              ; counter variable
ALPHA VAR WORD          ; counter variable
BETA VAR WORD           ; counter variable
STAT_1 VAR BYTE         ; status of led
STAT_2 VAR BYTE         ; status of led
```

(Continued)

Program 20.1 Final segment to make a finished program—the touch panel (*Continued*)

```
LCD_1 VAR BYTE            ; memory space
LCD_2 VAR BYTE            ; memory space
LCD_ONE VAR PORTA.2       ; LCD connection
LCD_ONE=0                 ; LCD turned off
LCD_TWO VAR PORTA.3       ; LCD connection
LCD_TWO=0                 ; LCD turned off
COL VAR BYTE              ; column counter
ROW VAR BYTE              ; row counter
TIM0 VAR WORD             ; time 0
TIM0 =10                  ;
TIM1 VAR WORD             ; time 1
TIM1=10                   ;
BLINK VAR BIT             ; blink status indication
BLINK=1                   ;
GLOW VAR BIT              ; glow status indication
GLOW=0                    ;
ACTIVE VAR BIT            ; active status indication
ACTIVE=1                  ;
INACTIVE VAR BIT          ; inactive status indication
INACTIVE=0                ;
                          ;
PAUSE 500                 ; pause for LCD startup
LCDOUT $FE, $01, "CLEARING LCD" ; clear LCD
PAUSE 250                 ;
LCDOUT $FE, $01           ;
                          ;
MAIN:                     ; main loop of the program
  GOSUB GETKEY            ; get the column and row
  GOSUB DISPLAY           ; display information on the LCD
  GOSUB DO_IT             ; take the necessary action
  COL=0                   ; rest the column memory
  ROW=0                   ; reset the row memory
GOTO MAIN                 ; do it forever
                          ;
GETKEY:                   ; routine to get the row and column
  PORTC=%11101111         ; turn on column 1
  PAUSE 1                 ; pause needed for LED to react
  IF PORTB.0=1 THEN COL=1 ; check photo for col 1 and save if is 1
    PORTC=%11011111       ; turn on column
    PAUSE 1               ; pause needed for LED to react
  IF PORTB.1=1 THEN COL=2 ; check photo for col 2 and save if is 1
    PORTC=%10111111       ; turn on column
    PAUSE 1               ; pause needed for LED to react
  IF PORTB.2=1 THEN COL=3 ; check photo for col 3 and save if is 1
    PORTC=%01111111       ; turn on column
    PAUSE 1               ; pause needed for LED to react
```

(*Continued*)

Program 20.1 Final segment to make a finished program—the touch panel (*Continued*)

```
   IF PORTB.3=1 THEN COL=4    ; check photo for col 4 and save if is 1
   PORTC=%11111110            ; turn on column
   PAUSE 2                    ; pause needed for LED to react
     IF PORTB.4=1 THEN ROW=1    ; check photo for row 1 and save
                                ; if is 1
     PORTC=%11110111          ; turn on column
   PAUSE 2                    ; pause needed for LED to react
   IF PORTB.5=1 THEN ROW=2    ; check photo for row 2 and save if
                              ; is 1
     PORTC=%11111111          ; turn everything off
RETURN                        ; end of routine
                              ;
DISPLAY:                      ; show selection and PORTB and C
LCDOUT $FE, $80,"R=",DEC1 ROW ," C=",DEC1 COL," D=",DEC3 TIM0,"_
D=",DEC3 TIM1
LCDOUT $FE, $C0, "B", BIN8 PORTB, " C", BIN8 PORTC
RETURN                        ;
                              ;
DO_IT:                        ; routine executes effect of selections
SELECT CASE ROW               ; first look at the rows
   CASE 1                     ; ROW 1
     SELECT CASE COL          ; look at the columns
                              ;
       CASE 1                 ; Column 1
         IF LCD_ONE=1 THEN    ; if LCD in ON
           LCD_ONE=0          ; turn if OFF
           LCD_1=INACTIVE     ; set it as inactive
         ELSE                 ; else
           LCD_1=ACTIVE       ; set it active
           LCD_ONE=1          ; turn it on
           STAT_1=GLOW        ; remember it is glowing not blinking
           TIM0=10            ; reset time to 10 count
         ENDIF                ; end of comparison
         GOSUB PAUSER         ; pause to de-bounce
                              ;
       CASE 2                 ; column 2
         IF LCD_1=ACTIVE THEN   ; react only if the LED is on
           IF STAT_1=GLOW THEN  ; if it is glowing
             STAT_1=BLINK     ; turn it to blinking
           ELSE               ; else
             STAT_1=GLOW      ; turn it to glow
             LCD_ONE=1        ; turn it on
           ENDIF              ; end of comparison
           GOSUB PAUSER       ; pause to de-bounce
         ENDIF                ; end of comparison
                              ;
```

(*Continued*)

Program 20.1 Final segment to make a finished program—the touch panel (*Continued*)

```
        CASE 3                  ; column 3
          TIM0=TIM0-1           ; decrease time to make faster blink
          IF TIM0 <1 THEN TIM0=1  ; if it is too low keep it at
                                ; minimum of 1
          GOSUB PAUSER1         ; delay for de bounce
                                ;
        CASE 4                  ; column 4
          TIM0=TIM0 + 1         ; increment the delay timer
          IF TIM0 >20 THEN TIM0=20 ; if it is too high, set it as
                                ; high
          GOSUB PAUSER1         ; delay for de-bounce
        CASE ELSE               ; here if there was something else
      END SELECT                ; end of selection of all the rows
                                ;
    CASE 2                      ; row 2
      SELECT CASE COL           ; look at the columns
        CASE 1                  ;
          IF LCD_TWO=1 THEN     ; the rest of the code duplicates what
            LCD_TWO=0           ; was done for row 1 above line
                                ; for line
            LCD_2=INACTIVE      ;
          ELSE                  ;
            LCD_2=ACTIVE        ;
            LCD_TWO=1           ;
            STAT_2=GLOW         ;
            TIM1=10             ;
          ENDIF                 ;
          GOSUB PAUSER          ;
                                ;
        CASE 2                  ;
          IF STAT_2=GLOW THEN        ;
            STAT_2=BLINK        ;
          ELSE                  ;
            STAT_2=GLOW         ;
          ENDIF                 ;
          GOSUB PAUSER          ;
                                ;
        CASE 3                  ;
          TIM1=TIM1-1           ;
          IF TIM1 <1 THEN TIM1=1  ;
          GOSUB PAUSER1         ;
                                ;
        CASE 4                  ;
          TIM1=TIM1+1           ;
          IF TIM1 >20 THEN TIM1=20         ;
          GOSUB PAUSER1         ;
        CASE ELSE               ;
```

(*Continued*)

Program 20.1 Final segment to make a finished program—the touch panel (*Continued*)

```
                              ;
    END SELECT                ;
  CASE ELSE                   ;
END SELECT                    ;
;The following segment of code uses the two "time counters" to
;turn the two LEDs on
;and off as specified by the faster/slower buttons. It counts each
;timing sequence
;independently and turns the related LED on and off.
IF LCD_1=ACTIVE THEN          ;
  IF STAT_1=BLINK THEN        ;
    ALPHA=ALPHA +1            ;
    IF ALPHA>=TIM0 THEN       ;
      ALPHA=0                 ;
      TOGGLE LCD_ONE          ;
    ENDIF                     ;
  ENDIF                       ;
ENDIF                         ;
IF LCD_2=ACTIVE THEN          ;
  IF STAT_2=BLINK THEN        ;
    BETA=BETA +1              ;
    IF BETA>=TIM1 THEN        ;
      BETA=0                  ;
      TOGGLE LCD_TWO          ;
    ENDIF                     ;
  ENDIF                       ;
ENDIF                         ;
RETURN                        ;
                              ;
PAUSER:                       ; pause to de-bounce, long
  FOR X=1 TO 300              ;
  PAUSE 1                     ;
  NEXT X                      ;
RETURN                        ;
                              ;
PAUSER1:                      ; pause to de-bounce, short
  FOR X=1 TO 50               ;
  PAUSE 1                     ;
  NEXT X                      ;
RETURN                        ;
END                           ;
```

EXERCISE

Modify the hardware and the Program 20.1 so the four columns are detected on lines B4 to B7, and then use the interrupt capability of these four lines to improve the performance of this program by making its operation crisper.

21

SINGLE SET POINT CONTROLLER WITH REMOTE INHIBIT CAPABILITY

Project 7

In this project, we take the dual temperature thermometer we have made and turn it into an adjustable programmable single-point controller that can control one property (not necessarily a temperature) in a process. The instrument can be controlled remotely. See Figure 21.1.

Any controller needs the following basic properties to be both useful and easy to use:

1. Have a set point that can be adjusted with ease
2. Have a detector for the property being measured
3. Have an output signal that is made available after 1 and 2 are compared
4. Have an input that can inhibit the operation of the controller from a remote location

Our dual temperature controller already has two inputs. One of these can easily be converted into a control point with the addition of a simple 5000 Ohms potentiometer. We also need to add a couple of connections for an output signal and we need to have a way to inhibit the operation of the system should we require it. This can be done with a simple Single Pole Single Throw (SPST) switch that can be located remotely wherever we want.

It can also be useful to have a switch that reverses the operation of the controller (in other words, that turns the outgoing signal *on* on an increasing rather than decreasing value, and vice versa). This is bought to your attention so you are aware of this feature, but it will not be undertaken on this project. It is simple enough to add a switch to do so.

THE CONTROL STATEMENT

"If the signal at the detector is less than (or more than) the control point, and if the inhibit signal is not CLOSED, then provide an output signal on the output line."

Figure 21.1 A single set point controller with inhibit. (When this board is mounted on a box, the temperature setter and the inhibit switch can be mounted on the box side or they can be mounted at remote locations.)

This is typical for most single-point controllers like a residential thermostat, a small hot water heater controller, or any number of similar on/off controllers. More importantly *having the "remote inhibit" capability allows this controller to be controlled by another instrument (remotely).*

Let's create the output on PORTA.5 and the inhibit signal on PORTA.3 because these are the two lines we have available on the board as it is. We have kept PORTA.4 free because this is the line that is used for some of the counter inputs in the PIC 16F877A. (In our case, PORTA.5 also just happens to be free.)

The additions to the hardware circuitry on the dual temperature sensor are shown in Figure 21.2.

In the software, we have to read the two analog signals, the set point and the controlled point, and make the necessary comparison. Then, based on the calculation, we need to turn the output signal on if the inhibit signal is not present. (Open in our case.)

The set point adjustment device is a potentiometer placed between 5 volts and ground. We read the wiper position of this potentiometer to get a reading between 0 and 255 as we have done many times before. This reading can be converted to the set point we need mathematically. (Remember that in the metronome project the 0–255 wiper position was converted to a number between 40 and 208.)

On the controller input, the variable we are reading can be any signal that can be converted into a resistance or a voltage. The conversion is to the equivalent of a 0- to 5-volt signal and is read into the microcontroller just as we did with the temperature readings. However, it does not have to be a temperature. What we are really

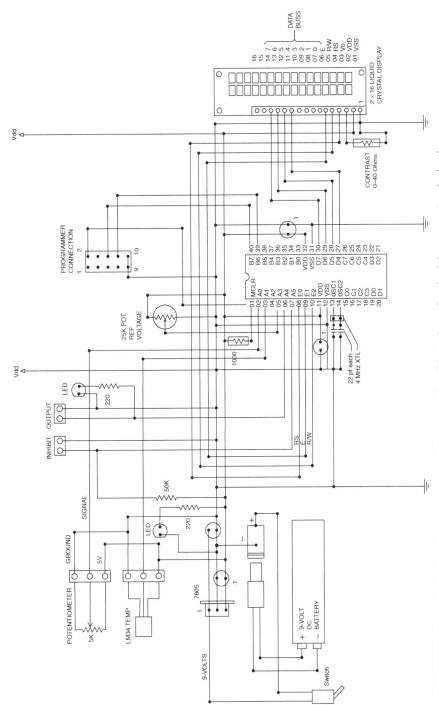

Figure 21.2 A single-point programmable controller. (Made from the dual thermometer device.)

reading is a voltage. Knowing the range of the signal we will be reading allows us to design a suitable electronic network to give us the range we need to feed into the microcontroller.

The input does not have to be a voltage or a resistance either. It can be a frequency that we can read with the COUNT and PULSIN commands. Similarly, the output does not have to be an on-off signal. It can just as easily be a pulse width modulated (PWM) signal or a frequency if that is desired.

With the preceding in mind, we will develop the thermostatic controller based on (1) the LM34 sensor because we already have the sensor and (2) discuss a thermistor-based controller because thermistors are inexpensive and give us an opportunity to read in a signal another way.

THE LM34-BASED CONTROLLER

The code development for using the LM34 is shown in Programs 21.1 and 21.2. The code uses the same nomenclature as was used for the dual thermostat instrument, but it modifies the code that was developed. The specific code for reading the two inputs does not have to be modified because in either case we are reading the equivalent of two potentiometers. We need to add code in the main loop to

1. Read the inhibit signal
2. Read the two inputs
3. Make the necessary comparisons
4. Output the result of the decision-making process as discussed earlier

Program 21.1   Inhibit code (Single-point programmable controller)

```
IF INHIBIT=1 THEN        ; if the inhibit switch is ON
   PORTB.1=0             ; turn OFF the signal
ELSE                     ; else
  IF VAL1 > VAL2 THEN    ; make comparison of the two values
     PORTB.1=0           ; turn OFF signal
  ELSE                   ; else
     PORTB.1=1           ; turn ON the signal
  ENDIF                  ; end of comparison
ENDIF                    ; end of inhibit comparison
```

THERMISTOR-BASED CONTROLLER CONSIDERATIONS

In order to use a thermistor, we need to add some hardware to make it easier to read the thermistor. The goal is to get a usable range of readings from the thermistor in the temperature range we are interested in. Thermistors with a high resistance and a high rate are easier to use because we need a minimum resistance of the total network to be 2k ohms or higher across the 5 volts we are using across the bridge.

The thermistor I used had the following properties:

Resistance at 70°F	1000 ohms
Delta	25 ohms per degree Fahrenheit in positive direction
Low temperature of interest	32°F
High temperature of interest	300°F

Under ideal conditions, the bridge we design would provide a signal close to 0 volts at 32° and close to 5 volts at 300°F. The network would have an overall (lower) resistance of about 5k ohms so we don't overload the power supply. A simple network is not going to be able to do this, but a usable network is possible.

From the preceding data we can calculate that the thermistor has about 50 ohms resistance at 32°F and about 6750 ohms at 300°F. If we put a 5000 ohms resistance in series with it, the thermistor can be connected to the PIC, as shown in Figure 21.3. The analog signal value we will read on the A-to-D conversion will be from 146 to 252.

$(50/5050)*255 = 2.52$ $255 - 3 = 252$ at 32°F

and $(6750/11750)*255 = 146.49$ $255 - 146 = 109$ at 300°F

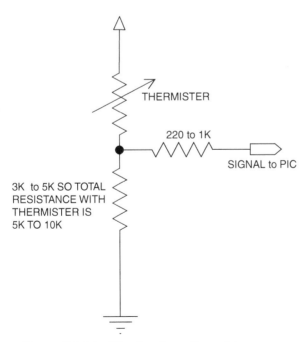

Figure 21.3 Circuitry for a thermistor connection to PIC. (Thermistor resistance can increase or decrease with temperature.)

As always, we need a resistance of about 5000 ohms in series with the thermistor so the circuit will not short across the power supply by drawing too much current when the thermistor is at its cold extreme.

In the usual case, the resistance in the circuit should be selected to match the ambient temperature resistance of the thermistor. This will give a reading of 127 or the wiper middle position at the ambient temperature. The resistance can be varied to raise and lower the ambient temperature reading. The total resistance of the resistance and the thermistor should be at least 5000 ohms for most applications.

The sensitivity of the readings will be best if the effect of the change in resistance of the thermistor is such that the range expected is approximately the same as the pot value; however, as mentioned earlier, the total resistance should not get below about 4k ohms to keep from overwhelming the power supply

Some other reliable method should be available to confirm that your instrument is providing accurate values for the temperatures being measured.

Other devices are used in a similar way. If a device provides a voltage, the voltage can be divided or amplified to be at a suitable level and then connected between ground and the input port pin. A reference voltage does not necessarily need to be connected to the PIC. Select and connect the Vdd source to suit.

READING THERMOCOUPLES

Thermocouples provide extremely low voltages and almost no current so we have to approximate the conditions provided by a Wheatstone bridge to read them. Alternatively, op amps that amplify the signal can be used. The industry provides a number of ICs that allow each of the types of thermocouples to be read into a PIC-type microcontroller with relative ease.

Program 21.2 Single-point controller: full program (This program runs on the board that fits on the box from All Electronics. See Figure 21.1 for an illustration.)

```
CLEAR                           ; clear all memory locations
DEFINE OSC 4                    ; system osc speed
DEFINE LCD_DREG PORTD           ; define LCD connections
DEFINE LCD_DBIT 4               ; data starting bit
DEFINE LCD_BITS 4               ; number of data bits
DEFINE LCD_RSREG PORTE          ; select register port
DEFINE LCD_RSBIT 0              ; select register bit
DEFINE LCD_EREG PORTE           ; enable register
DEFINE LCD_EBIT 1               ; enable bit
DEFINE LCD_LINES 2              ; lines in display
DEFINE LCD_COMMANDUS 2000       ; delay in micro seconds
DEFINE LCD_DATAUS 50            ; delay in micro seconds
LOW PORTE.2                     ; puts LCD in write only mode
DEFINE ADC_BITS 8               ; set number of bits in result
DEFINE ADC_CLOCK 3              ; set clock source (3=rc)
DEFINE ADC_SAMPLEUS 50          ; set sampling time in us
VAL0 VAR BYTE                   ; create to store result
VAL1 VAR WORD                   ; create to store result
```

(Continued)

Program 21.2 Single-point controller: full program (This program runs on the board that fits on the box from All Electronics. See Figure 21.1 for an illustration.) (*Continued*)

```
TRISA=%00111111                              ; set PORTA
TRISB=%00001000                              ; set PORTB
TRISD=%00000000                              ; set PORTD
TRIS =%00000000                              ; set PORTE
ADCON1=%00000010                             ; set analog pin selections
                                             ;
PAUSE 500                                    ; pause to start up LCD
LCDOUT $FE, $01, "CLEAR"                     ; display Clear message
PAUSE 500                                    ; pause to see message
LCDOUT $FE, $01                              ; clear the screen
PORTB.2=0                                    ; ground this
OPTION_REG.7=0                               ; pull all PORTB inputs high
LOOP:                                        ; main loop
ADCIN 0, VAL0                                ; read channel 0 potentiometer
ADCIN 3, VAL1                                ; read channel 3 temp
VAL1=10*VAL1/5                               ; calculate VAL1
LCDOUT $FE,$80,"TMP=",DEC3 VAL1 ; display information
LCDOUT $FE,$C0,"SET=",DEC3 VAL0 ; display information
IF VAL1>=VAL0 AND PORTB.3=1 THEN ; compare to setting
    PORTD.3=0                                ; set PORTD.3=0
ELSE                                         ; or
   PORTD.3=1                                 ; set PORTD.3=1
ENDIF                                        ; end decision
IF PORTB.3=0 THEN                            ; see if inhibit is on.
    PORTD.2=0                                ; set PORTD.2=0
ELSE                                         ; or
   PORTD.2=1                                 ; set PORTD.2=0
ENDIF                                        ; end decision
PAUSE 10                                     ; delay 0.01 seconds
GOTO LOOP                                    ; go back to loop and repeat
                                             ; operation
END                                          ; end of program
```

22

LOGGING DATA FROM A SOLAR COLLECTOR

Project 8

This project covers the automation of long-term data collection using a small and simple solar collector as the source of the data (see Figure 22.1).

There is a constant need for automated data logging because there are many times when we need instruments and controllers that can gather information unattended over long periods of time and simultaneously control a related function or two. We learn how to do this project with a solar data collector, which is the subject of this chapter.

> Dr. Sun of Peking Polytechnic University in Beijing, China is running a project for the United Nations to determine how much energy is available from the sun on a yearly basis at every location around the world. He has asked the chairman of our electronics department to look into how you and your fellow students might help in this endeavor. A very simple and inexpensive way to collect the data automatically is desired. Many thousands of units will be needed and the budget is limited, so he is not looking for the most accurate instruments in the world but has stressed that he does want reliability. This investigation is being made as a direct result of Dr. Sun's request.

In this exercise, we will build a small solar collector that will let us collect valuable data over an extended period of time automatically. The solar collector will have a ventilation fan built into it to reduce the temperature within the collector when it gets past a certain temperature, so as to more accurately reflect the collection conditions in a collector. The fan will be controlled by our controller. We will also keep track of the (inside) temperature of the air moved by the fan so we can determine how much energy the collector is ejecting. We will not actually measure how much air the fan moves. Instead, Dr. Sun's team will make an estimate of that during the analysis phase of the work.

Figure 22.1 A solar collector for a data logging project.

Figure 22.2 shows a single-line schematic of the collector with the sensors and other ancillary equipment for initial review as we proceed with the discussion.

The efficiency of a "hot air" solar collector decreases as the temperature inside it increases. In temperate climates, the maximum temperature can rise above 400°F in a commercial solar collector. If the energy is not removed from the collector, no more energy

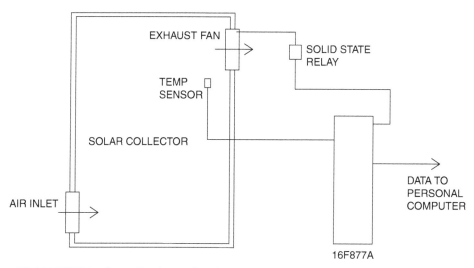

Figure 22.2 Overall schematic of our solar collector. (See the wiring diagram in Figure 22.3 for further details.)

will be collected. The collector ceases to collect further energy for a number of reasons that do not concern us here. Our interest is focused on how to remove energy from the collector so we can maximize the amount of energy that can be collected. We can do this by adding a small cooling fan, like one recycled from an old computer cabinet, to move the air through the collector. We could also add a couple of fabric check valves at the inlet and outlet to keep the air from blowing backwards through the collector when the fan is off, if necessary. We will use 135°F always (adj) as the max permitted temperature in our experiment to reflect residential heating conditions. We will turn the fan on if the internal temperature reaches 135°F but this temperature can be made adjustable by comparing it to the reading on one of the onboard potentiometers (the set point).

We will collect the data once every minute to get an accurate profile of how much energy the collector captures during the day. Let's assign 1 byte to each digit we have to collect, even though we are aware that the data could be packed much more tightly. The temperature will be read in degrees Fahrenheit so we will need 3 bytes for the temperature. The information we collect consists of the following data points:

1. Date and time of day, 12 bytes (YYMMDDHHMMSS)
2. Temperature inside the cabinet, 3 bytes (TTT)
3. Whether the fan is on or off, 1 byte (F = 0 or 1)

These 16 bytes have to be collected 60 times an hour all day and all night. (We are collecting nighttime data to allow Dr. Sun to estimate the heating needs in the area during evenings.) We have to collect an average of 23,040 (16 × 60 × 24) bytes every day. To make things move along and to allow for overhead and so on, let's say that we need space for 25K of data every day in round numbers, which would be 9125K of data in a year. Not a large amount of data considering the capability of even the smallest computers on the market, and within reach of storage in an MCU-based engine without any external storage if we could add a modest amount of memory to the MCU. We could ship the data to Dr. Sun at the Peking all these also need to be changed to keep with Beijing elsewhere Polytechnic University in Beijing on a CD-ROM or over the Internet, if the collection point has Web access. (We should contact Dr. Sun to inform him that the project could very easily be modified to collect the data every 30 seconds, or even more often, at no additional cost.)

Our preceding calculations have indicated that the problem is completely within our reach and so we can proceed with the fabrication of our collector and the design of the hardware and software we will need to get the job done. The drawings for the collector are on the Web site that supports this text, so we will not go into fabrication details here. Our interest is in the electronics and the software, but you are encouraged to make up a rudimentary collector to see how the project comes together. A few sheets of cardboard, a bit of sheet Styrofoam, a few scraps of wood, and some plastic sheeting will make a surprisingly effective collector that you can experiment with.

For the complete project, we need the items shown in Table 22.1:

Study the wiring diagram in Figure 22.3 to get familiar with how the various components are wired to the LAB-X1 before starting on the fabrication.

TABLE 22.1 LIST OF MATERIALS NEEDED FOR SOLAR COLLECTOR CONSTRUCTION

1. LAB-X1 board
2. A clock chip: NJU 6355
3. A small fan that runs at 120 volts
4. Glass or plastic to fit
5. One LM34 temperature sensor
6. One old IBM-PC, working
7. One hard drive with 10 MB of memory free
8. One serial cable for COM1
9. One 2-feet-by-8-feet sheet of 1-inch Styrofoam
10. Half a sheet of ½-inch exterior grade plywood
11. Some 1-inch brads
12. Some wood glue
13. The usual odds and ends

Microcontroller Hardware

Since it would be a major expense and effort for each student to design and build a controller for this project, and since we already have a LAB-X1 available to us, we will use the LAB-X1 as the controller for the project. This will limit what we can do a bit but since the main function of the exercise is data collection, this will not distract from the experience we are seeking.

In order to use the LAB-X1, we need to make the following changes/modifications to the board:

1. Add a clock chip in U6.
2. Add a crystal for the clock chip at J5.
3. Add headers to the LAB-X1 at P1, P2, P3, and P4 so we can connect to the MCU as needed.

The diagram in Figure 22.2 reflects the modifications that have to be made. Of particular interest is the fact that there are three potentiometers on the board that can interfere with our using the three lines that they are connected to for reading any inputs. However, the potentiometers can be used to modify variables we are using (for example, the temperature at which we turn on the fan), and we can still use the lines for I/O if we set the potentiometers to appropriate intermediate positions. If we do that, the inputs will be able to override the intermediate settings and we can read the input. (The program as written does not use these pots, but you should be aware that this can be done.)

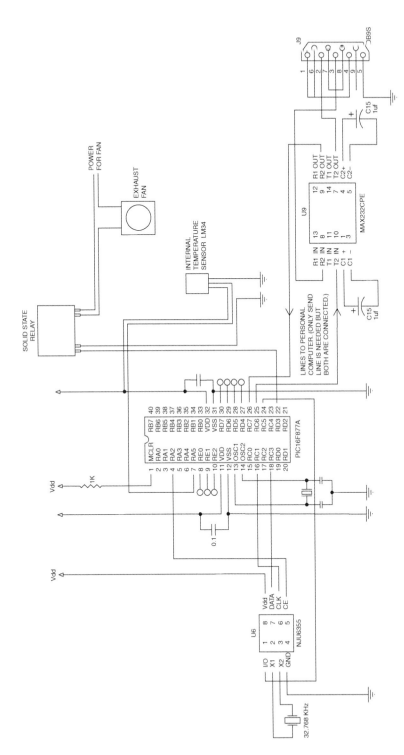

Figure 22.3 Wiring diagram for the solar collector instrument/controller/data transmitter.

Software

Note *The software listings are written as if the controller was a LAB-X1 board so that a new board does not have to be created to execute the requirements of the experiment. A much more versatile system could be created if a new board was designed. Since the major problem with the LAB-X1 is the fact that there are three pots already attached to the analog input lines, the circuit layout for this experiment must reflect this.*

The devices we will be using are to be attached as follows:

The temperature will be read from PORTA.5.

The clock uses the NJU6355 chip and is located as shown on the LAB-X1 schematic at U6.

A 32.786-KHz crystal must be added to the board at J5 to support the clock.

The fan switch will be at PORTD.3.

The interface to the computer will be the DB9 connector on the board.

The LCD display addresses will be the standard LCD addresses for the board.

The software for the solar collector has to perform the following functions:

Provide a clock function (with the hardware addition)

Read the temperature sensor

Make minor decisions / Control the fan

Send the data to the computer periodically

Recover from power outages automatically

The hardware needed to support these functions is shown on the wiring diagram in Figure 22.2, with all the pins that will be used identified.

If there is no battery backup, the clock must have an arbitrary starting time so the recovery from a power outage can be effected based on the data collected during the previous and following day. When the data is analyzed, any reset/discontinuity of the clock will indicate a power outage and the readings will have to be adjusted for time of day based on extrapolation from the daily cycles. Dr. Sun has indicated that these functions are to be performed when the data is finally analyzed by his group in Beijing.

Let's write the software as subroutines, one for each function. Each of the subroutines needs to have some preliminary work done up front during the setup of the program, and this is listed separately before each subroutine.

First, let's set up the LCD as we always do. Then, we can call the code out as a subroutine in the program where we need it. See Program 22.1.

Program 22.1 Data logging

```
CLEAR                           ; clear the RAM
DEFINE OSC 4                    ; define Osc Speed
DEFINE LCD_DREG PORTD           ; define the LCD connections
DEFINE LCD_DBIT 4               ; as we always do
DEFINE LCD_RSREG PORTE          ;
DEFINE LCD_RSBIT 0              ;
DEFINE LCD_EREG PORTE           ;
DEFINE LCD_EBIT 1               ;
LOW PORTE.2                     ;
                                ;
TRISA = %00010000               ; set PORTA
TRISB = %11110000               ; set PORTB
TRISC = %11110000               ; set PORTC
TRISD = %00000000               ; set PORTD
TRISE = %00000000               ; set PORTE
ADCON1= %00000111               ; don't forget to set ADCON1
                                ;
PAUSE 500                       ; pause .500 second for LCD startup
LCDOUT $FE, 1, "Clear"          ; clear LCD, go to first line
                                ;
MAIN:                           ;
GOSUB READ_CLOCK                ;
GOSUB READ_SENSORS              ;
GOSUB CONTROL_FAN               ;
GOSUB UPDATE_LCD                ;
IF RTCSEC =$00 THEN GOSUB SEND DATA ; send data when seconds read 00
GOTO MAIN                       ;
END                             ;
```

The subroutines called in the preceding code are developed in the following section. A short description precedes each code listing.

SETTING THE INTERNAL CLOCK

There are seven registers of interest in the NJU6355 clock module. They contain the following pieces of information:

Year

Month

Date

Day

Hours

Minutes

Seconds

As programmed for this experiment, the clock is set to 01 Jan, 2007 on startup, and after every power failure. It can, of course, be set to any date and time you choose, and it will not affect the results of our experiments in that we are not using the exact time/date information as such, but rather the daily cycle of the collected data. Even missing a day or two of data would not adversely affect our effort.

We have no way of intervening to set the clock if/when there is a power failure, so we set the clock to the beginning of the year 2007 on startup and whenever there is a power failure. Later on, when the data is analyzed, we can use the temperature data and number of data points since the last night cycle to determine the approximate time of day for each recovery. Fortunately, the exact time at which the readings are taken is not important in this particular investigation.

Note *There is a provision for a battery backup for the clock on the LAB-X1, and a battery could be provided to keep the clock going during power failures if this was desired. However, chances are, the batteries will not be easily available in many parts of the less-developed world where data must be gathered, so we should take that into consideration.*

The "date data" does not show up on the display but is sent to the computer along with the other data as a matter of course during each transmission to the PC.

The clock is seen as a set of memory locations by the MCU. The actual locations will depend on the clock chip we select. For this application, we have selected the NJU6355 IC chip. The use of this chip is explained in some detail in Chapter 7. We read the clock but the exact actual time is not of interest because we do not have a way to reset to the correct time after a power failure. Our interest is in the time that has elapsed since the last sunrise and since the last power failure. We can estimate the actual time from the information gathered for the day if we know the first date that the data was collected. (The program segment to read the IC is taken from a program on the microEngineering Labs Web site.)

The routine to send the data to the computer looks at the seconds value, and then every minute, based on the changes in the registers, it sends the relevant information to the computer for storage.

The initialization code for the clock module is shown in Program 22.2.

Program 22.2 Writing to the clock

```
                        ; The alias pins are as follows
CE VAR PORTA.2          ;
CLK VAR PORTC.1         ;
SDATA VAR PORTC.3       ;
IO VAR PORTC.5          ;
                        ; allocate variables
RTCYEAR VAR BYTE        ;
RTCMONTH VAR BYTE       ;
RTCDATE VAR BYTE        ;
RTCDAY VAR BYTE         ;
RTCHR VAR BYTE          ;
RTCMIN VAR BYTE         ;
RTCSEC VAR BYTE         ; set variables
```

(Continued)

Program 22.2 Writing to the clock (*Continued*)

```
LOW CE                   ; disable RTC
LOW CLK                  ;
HIGH IO                  ;
RTCYEAR=$07              ;
RTCMONT=$01              ;
RTCDATE=$01              ;
RTCDAY=$01               ;
RTCHR=$00                ;
RTCMIN=$00               ;
RTCSEC=$00               ;
IO=1                     ; set RTC to input
CE=1                     ; enable transfer
                         ; Write to all 7 RTC registers,
                         ; this is a reset condition
SHIFTOUT SDATA, CLK, LSBFIRST, [RTCYEAR, RTCMONTH, RTCDATE,_
RTCDAY\4, RTCHR, RTCMIN, RTCSEC]
```

Program 22.3 Reading the clock

```
READ_CLOCK:              ; clock reading subroutine
IO=0                     ; set RTC to output
CE=1                     ; enable transfer
                         ; read all 7 RTC registers
SHIFTIN SDATA, CLK, LSBPRE, [RTCYEAR, RTCMONTH, RTCDATE, RTCDAY\4,_
RTCHR, RTCMIN, RTCSEC]   ;
RETURN                   ;
```

On the LAB-X1, there are seven lines, distributed across ports A and E, that can be used for analog inputs. Lines A.0, A.1, and A.3 are connected to the three pots on the card and so cannot be used. Line A.2 is being used by the real-time clock, and line A.4 is not available for analog input (open collector). *This leaves only line A.5 as a free analog input line* (this is analog channel 4). Lines on port E are being used to control the display and are thus not available to us. Accordingly, we will use the channel 4 line for reading the internal temperature of the collector.

The initialization code for reading the sensor is as follows:

```
                         ; define the A2D set up
DEFINE ADC_BITS 8        ; set number of bits in result
DEFINE ADC_CLOCK 3       ; set internal clock source (3=rc)
DEFINE ADC_SAMPLEUS 50   ; set sampling time in uS
                         ; define variables used
INTEMP VAR BYTE
```

READ THE SENSORS

The routine in Program 22.4 reads each sensor and stores its value in its appropriate variable. The Display Routine will read these values from the variables and put them in the display when necessary.

Program 22.4 Reading the sensors

```
READ_SENSORS:    ;
ADCIN 4, INTEMP  ; read channel 4 into INTEMP if one of the pots
                 ; is to modify this temp, add the code to read
                 ; it in here.
RETURN           ;
```

CONTROLLING THE FAN

This routine looks at the internal temperatures, and if the temperature is above 135°F, it turns the fan on. 135°F approximates the temperature of the air coming out of a residential furnace in winter. See Program 22.5.

We will use pin D.3 to control the solid state relay so we can see the condition of the relay on the bargraph when the relay is turned on. Some relays have an LED on them, some don't. We will be covered in either case.

Program 22.5 Controlling the fan

```
CONTROL_FAN:          ;
IF INTEMP >=135 THEN  ; if one of the pots is to modify this temp,
                      ; add that logic here on this line.
PORTD.3=1             ;
ELSE                  ;
PORTD.3=0             ;
ENDIF                 ;
RETURN                ;
```

UPDATING THE LCD

During each 1-second cycle, the system reads all the variables to appear on the LCD and displays them. It turns the fan on and off as necessary. The display is a convenience for the operation of the system. It can accommodate two lines of 20 characters each and we will fill these in as follows:

D=YY:MM:DD:HH:MM:SS

TEMP=TTT   FAN=X

This arrangement allows the investigator to look at current conditions at any time. If the investigator needs to know the actual time, they can look at a wrist watch. Program 22.6 implements the updating of the LCD.

Program 22.6 Updating the LCD

```
UPDATE_LCD: ;
LCDOUT $FE  $80, "D=", HEX2 RTCYEAR, ":", HEX2 RTCMONTH, ":", HEX2_
RTCDATE," ",HEX2 RTCHR, ":", HEX2 RTCMIN, ":", HEX2 RTCSEC_
LCDOUT $FE  $C0, "TEMP=",DEC3 INTEMP,"  FAN=",DEC1 PORTD.3
RETURN      ;
```

SENDING THE DATA TO A COMPUTER

The routine in Program 22.7 checks the time on the internal clock to determine when to send the data to the computer. It monitors the seconds display. When 1 minute has passed, the seconds go to 00. At that time, the program jumps to the Send routine and all the data to be recorded is sent to the computer for storage. Then, it pauses for 1 second to make sure it will not see 00 in the seconds counter again within this minute and continues.

Program 22.7 Sending data to the computer

```
SEND_DATA:                                          ;
LCDOUT $FE, $80, "SENDING DATA       "              ;
LCDOUT $FE, $C0,"                    "              ; clears the line
SEROUT PORTC.6, T2400, [RTCYEAR, RTCMONTH, RTCDATE, RTCHR, RTCMIN,_
RTCSEC, INTEMP, PORTD.3, 10, 13]
ELSE                                                ;
ENDIF                                               ;
PAUSE 1000                                          ;
RETURN                                              ;
```

If there has been a power failure, the appropriate compensation will be made when the data is analyzed. If for some reason the power is off for more than a day, there will be a problem with data lost for that day, but for an investigation of this kind that is not really critical.

When we combine all of the preceding routines in one listing, we get what's shown in Program 22.8. This program is set up to record the data every 5 seconds so you can see what is going on both in the LAB-X1 and in the computer in real time. The computer must be set up as a dumb terminal to match the requirements of the output from the LAB-X1 for all this to work as desired. After you are sure the program is working as expected, you can change the time between transmissions to once a minute or whatever is desired.

Program 22.8 The finished program for the solar collector (Solar collector–based data logging)

```
CLEAR                              ; clear the RAM
DEFINE OSC 4                       ; define osc speed
INCLUDE "MODEDEFS.BAS"             ; include shiftin/out modes
DEFINE LCD_DREG PORTD              ; define the LCD connections
DEFINE LCD_DBIT 4                  ;
DEFINE LCD_RSREG PORTE             ;
DEFINE LCD_RSBIT 0                 ;
DEFINE LCD_EREG PORTE              ;
DEFINE LCD_EBIT 1                  ;
DEFINE ADC_BITS 8                  ; set number of bits in result
DEFINE ADC_CLOCK 3                 ; set internal clock source (3=rC)
DEFINE ADC_SAMPLEUS 50             ; set sampling time in us
LOW PORTE.2                        ;
                                   ;
```

(Continued)

Program 22.8 The finished program for the solar collector (Solar collector–based data logging) (*Continued*)

```
TRISA= %00111111                 ; set PORTA
TRISB= %00000000                 ; set PORTB
TRISC= %00000000                 ; set PORTC
TRISD= %00000000                 ; set PORTD
TRISE= %00000000                 ; set PORTE
ADCON1= %00000111                ; don't forget to set ADCON1
                                 ; alias pins are as follows
CE VAR PORTA.2                   ; Real time clock mode
CLK VAR PORTC.1                  ;
SDATA VAR PORTC.3                ;
IO VAR PORTC.5                   ;
; allocate variables             ;
RTCYEAR VAR BYTE                 ; year
RTCMONTH VAR BYTE                ; month
RTCDATE VAR BYTE                 ; date
RTCDAY VAR BYTE                  ; day
RTCHR VAR BYTE                   ; hour
RTCMIN VAR BYTE                  ; minute
RTCSEC VAR BYTE                  ; seconds
INTEMP VAR BYTE                  ; temperature
                                 ;
LOW CE                           ; disable RTC
LOW CLK                          ;
HIGH IO                          ;
ADCON1 = 7                       ; PORTA and e digital
LOW PORTE.2                      ; lcd r/w low = write
; set initial time to 00:00:00 am on 01/01/07 same will be true
; for all resets.
RTCYEAR  = $07                   ;
RTCMONTH = $01                   ;
RTCDATE = $01                    ;
RTCDAY= 2                        ;
RTCHR = 0                        ;
RTCMIN= 0                        ;
RTCSEC= 0                        ;
IO = 1                           ; set RTC to input
CE = 1                           ; enable transfer
; write to the 7 RTC registers to initialize them
SHIFTOUT SDATA, CLK, LSBFIRST, [RTCYEAR, RTCMONTH, RTCDATE,_
RTCDAY\4, RTCHR, RTCMIN]
CE = 0                           ; disable RTC
PAUSE 500                        ; pause .500 second for LCD
LCDOUT $FE, 1, "CLEAR"           ; clear LCD
                                 ;
MAIN:                            ;
GOSUB READ_CLOCK                 ;
GOSUB READ_SENSORS               ;
```

(*Continued*)

Program 22.8 The finished program for the solar collector (Solar collector–based data logging) (*Continued*)

```
GOSUB CONTROL_FAN             ;
GOSUB UPDATE_LCD              ;
GOSUB SEND_DATA               ;
GOTO MAIN                     ;
                              ;
READ_CLOCK:                   ;
IO= 0                         ; set RTC to output
CE= 1                         ; enable transfer
                              ; read all 7 RTC registers
SHIFTIN SDATA, CLK, LSBPRE, [RTCYEAR, RTCMONTH, RTCDATE, RTCDAY\4,_
RTCHR, RTCMIN, RTCSEC]
CE = 0                        ; disable RTC
RETURN                        ;
                              ;
READ_SENSORS:                 ;
ADCIN 4, INTEMP               ; read channel 4 to intemp
RETURN                        ;
                              ;
CONTROL_FAN:                  ;
IF INTEMP=>135 THEN           ;
PORTD.3=1                     ;
ELSE                          ;
PORTD.3=0                     ;
ENDIF                         ;
RETURN                        ;
                              ;
UPDATE_LCD:                   ;
LCDOUT $FE, $80, "TIME=",HEX2 RTCHR, ":", HEX2 RTCMIN, ":", HEX2_
RTCSEC," "                    ;
LCDOUT $FE, $C0, "TEMP=",DEC3 INTEMP,"  FAN=",DEC1 PORTD.3
PAUSE 10
RETURN                        ;
                              ;
SEND_DATA:                    ;
IF RTCSEC=$00 THEN            ;
LCDOUT $FE ,$80 ,"SENDING DATA" ;
LCDOUT $FE, $C0, "              ";
; in the next line we decide what we are going to
; send to the computer for storage. Day of week is omitted.
SEROUT PORTC.6, T2400, [RTCYEAR, RTCMONTH, RTCDATE, RTCHR,_
RTCMIN, RTCSEC, INTEMP, PORTD.3, 10, 13]
    PAUSE 1100                ; so we are into the next second.
    ELSE                      ; do nothing
    ENDIF
    RETURN                    ;
                              ;
    END                       ;
```

In Program 22.8, the data is stored on disk in hexadecimal format. If you want it in decimal format, you have to read it into a program variable and then convert it to the desired format. As provided (in the compiler), the SEROUT instruction offers a limited choice for the data transfer as it was read in from the clock.

Note *There is a certain amount of nonvolatile memory on the PIC 16F877A that could have been utilized to store power failure information for future analysis. However, because there is no provision that allows us to reset the clock accurately, the use of this memory cannot provide a more perfect solution.*

PROBLEM!

When this program was sent to Dr. Sun, one of his students complained that the program was not completely correct. Can you find what he or she thinks the problem is? The answer is provided at the end of the support Web site.

23
DEBUGGING

General

It takes a long time to become a good debugger. It has taken me forever, and I am still learning and still not very good at it. The problem is that you must have a good understanding of the PIC you are using and come up with a solid design. Though this may seem pretty straightforward, it isn't, especially if you are using a new PIC—and the PIC 16F877A we are discussing is new to most of us.

In plain English, this means that I really do not know how to tell you to become a good debugger. Even so, I need to give you some guidance to help you, so I have put down everything that came to mind as I was debugging the various projects in this book. In these paragraphs, I have allowed myself some repetition under the various headings so they can each stand alone.

At the end of the chapter, I have included some questions and answers that were the result of some discussions on the Web site for another project but that are also relevant to any discussion about the 16F877A PIC.

Debugging and Troubleshooting

Debugging is not a random process during which one might hope to get lucky. It is a very carefully thought out strategy to find out why a program is not behaving the way it was intended to and what needs to be done to correct the problem. *You will have fixed the problem only if you can make the problem come back by undoing the fix.* A vague suspicion that you might have fixed the problem by pressing on a warm component is not enough to conclude that the problem has actually been solved. You must be able to make the problem come back. This is exactly the reason why intermittent problems are so hard to fix. It's hard to make them come and go on command. In other words, it is harder to clearly understand the problem because it is intermittent.

First Problem that Must Be Fixed: The Microcontroller Crystal Must Oscillate

If the PIC oscillator will not oscillate, nothing can be done to fix anything, so the first thing we must do is make sure the OSC lines are actually oscillating. The easiest way to check this is with an oscilloscope. If there are problems, the following points are relevant.

1. Remember that the LAB-X1 runs at 4 MHz out of the box. Your system should also be designed to run at 4 MHz so you can use the LAB-X1 as a test bed whenever you need to. This allows you to take the PIC back and forth between your project and the LAB-X1 to see where the problems are. This is *not* a trivial tool that we have at our casual disposal.
2. To start with, do all your projects with a 4 MHz crystal. Later, you can move to 20 MHz. It is easy enough to replace a crystal on a board.
3. If the crystal frequency does not match the DEFINE OSC statement in your program, there can/will be problems. If you are using a crystal or resonator, you must know what the frequency of the device is and your software must state this number in the OSC statement at the top of the program accurately. They must match.
4. The Configuration and Option pulldown menus in the programmer must match the actual conditions in the hardware and they must match each other.

START WITH THE FOLLOWING CHECKS ON THE HARDWARE SIDE

1. Make sure the microcontroller has power.
   Make sure it is 5 volts on the money.
   Make sure there is power to both sides of the microcontroller, as is required on a large number of them.
   Make sure ground is a solid ground at all points in the circuitry. If the ground is squirrelly, this is not good. Fix it.
2. Make sure the MCLR pin has been pulled up to 5 volts with a 1K to 10K resistor. This value depends on the design you are using. Make sure the MCLR pin is actually high. Put a meter on it. Use an oscilloscope if you suspect transient operation.
3. Make sure the oscillator is running. Use an oscilloscope. Make sure the operation is consistent.
4. Use your eyes to check the PC board for shorts and dry solder joints. Use a magnifying glass in difficult areas. Go over questionable areas with a soldering iron again and recheck your work.
5. Make sure the wiring is what you think it is. Check the route of every wire. Mark it off on the schematic as you check it. Check the PC board trace routings where necessary.
6. Check and confirm the values of each of the components on the board.

7. Make sure each IC is oriented with pin 1 in the proper location in its socket.
8. Make sure power and ground to each IC are properly routed and are actually at the values they are supposed to be with appropriate instruments.
9. Measure voltages throughout the layout and confirm they are what they are supposed to be.
10. Make sure all capacitors are installed correctly. Check polarity where necessary.
11. Make sure all diodes are installed correctly. All of them don't always need to be installed so the cathode is connected to ground. Confirm connections on all inductive loads.
12. Make sure all inductive (back electro motive force [EMF]) loads are properly protected against with diodes. Make sure these diodes are properly rated for amperage, voltage, and switching speed.
13. Use the oscilloscope to check for noise in the circuitry. Eliminate it by adding small capacitors to ground at the noisy areas.
14. Read the relevant chapter of the datasheet (on startup) again. Make sure you understand what the datasheet is saying.
15. Make sure the processor has not been destroyed. Sometimes one or more lines on various ports can be destroyed. Try a replacement unit.

MAKE THE FOLLOWING CHECKS ON THE SOFTWARE SIDE

1. Go over the software line by line and make sure there are no typos. Not all typing mistakes are identified by the compiler software.
2. Write a short LED blink routine and run that on the board to make sure the system is actually alive and working. If the program loops as part of its design, add a blinking instruction for one of the LEDs to tell you that the loop is actually executing as designed and not hanging up on some other segment of code.
3. Use MicroCode Studio to see what is going on in the system as it is run. The software is free to download off the Internet for individual users and will run the 16F877A. Details on using this software are beyond the scope of this book and are not covered here, but it's worth the effort to learn this software.
4. For the preceding step, you can set up your PC as a dumb terminal and connect it to the board so the entire system can be viewed on the screen with the appropriate software.
5. Go over the software to make sure there are no logical errors.
6. Follow the use of each variable throughout the program to make sure it does not get modified where it should not.
7. The software we are using employs integer math and 8- and 16-bit variables. Make sure that none of the variables exceed the bounds that have been designated by the variable sizes. Be sure you understand what happens when registers overflow or underflow in integer math. You cannot use decimal points at most locations in integer math unless you take special precautions.
8. Read the "microprocessor start-up" chapter (Chapter 14) of the datasheet again. Make sure you understand what the datasheet is saying.

There are a number of ways to *get input into and feedback out* from a malfunctioning program. Using as many as possible will reduce the number of times you have to reload the program. You have the following feedback devices available to you. Incorporate what you need into critical areas of the program.

1. The LCD—use both line 1 and line 2, and use each character on each line.
2. The speaker—it's simple to set up two tones that are easy to differentiate.
3. The dumb terminal program for a PC can display a large amount of information.
4. Use the various buttons to modify what is happening in the program in real time.
5. Use the three onboard potentiometers to input various values into the program and see how they modify its operation.
6. If more is needed, add LEDs to the hardware so you can turn them on and off at critical junctures.
7. Use the DEBUG command in PBP to send information to a dumb terminal. The use of this command is explained in detail in the PBP manual. See the following also.

   You can *insert a short loop* that displays the registers you are interested in at a critical location in the program. When the program enters this loop, it indicates that the program actually got this far and then it displays the registers of interest again and again without going any further. This loop can be moved up and down through the program to see what is going on where.

8. Determine if the program is actually getting to a certain critical line of code.
9. Determine what the contents of various registers are at critical times in the code.
10. Look at how counters are behaving to confirm that this is exactly what is supposed to be happening.
11. Display data based on interrupts programmed and/or entered by you from the keys.
12. Look for areas where the program might be getting stuck in a loop.
13. Pay special attention to the handling of interrupts.
14. Go over the circuit layout to make sure there are no mistakes in the design of the circuitry.
15. Go over the physical circuit to make sure it is actually wired the way it was designed to be.
16. Make sure all lines that are to be pulled up or down are actually being pulled up and down and that the resistors being used are of the right values.

If the Chip Refuses to Run

If the chip refuses to run at all, check the configuration settings in the programmer software. The oscillator configuration is the most critical, but other settings can also prevent the PICmicro from starting up. See the "Special Features of the CPU" chapter in the datasheet (Chapter 14) for correct configuration details. The default settings for configuring the various conditions are discussed later near the end of this chapter.

Using the PBP Compiler Commands to Help Debug a Program

The PICBASIC PRO Compiler provides a number of commands that can be a tremendous help in debugging programs that refuse to cooperate. These commands can be broken down into a number of categories to better understand how they can be used.

Commands that Can Provide Debug Output to a Serial Port

A number of commands provided by the compiler do not have a function other than to aid in the debugging of programs by outputting data to a designated pin on the microcontroller. This data can then also be displayed on a dumb terminal.

The following commands are useful in the context. See the PBP manual for details.

DEBUG is like a print command to the serial port (as well as to the dumb terminal)

DEBUGIN

ENABLE

ENABLE DEBUG

DISABLE

DISABLE DEBUG

PEEK

POKE

SOUND

Dumb Terminal Programs

A number of terminal programs are available at no charge on the Internet. I use the dumb terminal program provided by Microsoft as a part of their operating system utilities. It provides all the functionality you need to use it with the PICBASIC PRO Compiler and MicroCode development software.

The Bray terminal program is a more sophisticated dumb terminal program and is available for free on the Internet.

Solderless Breadboards

Using solderless boards for your prototyping activities is, in general, not recommended. They are fine for small experimental excursions when you first start out with the microcontrollers, but as your circuitry gets more and more complicated, there is too great chance for poor connections and wires to come loose to use these devices.

It is recommended that you use the perforated boards that have a separate solder pad for each hole and solder each component into the board. Then, wire each of the components with hookup or wire wrap wire with straightforward point-to-point wiring. The key is to be very careful and thorough so there are no mistakes. It takes patience and care. Take your time and check your work before and after each connection is made.

(I also use circuit boards with continuous bars of conductors on them.)

Debugging at the Practical Level

Help! What do I do now? The project is "deader than a door nail" and I don't have a clue!

A fairly long program that you wrote will not work the way it is supposed to. You don't know if it is the software or the hardware and you do not have a clue as to what you should do. Don't throw it all in the garbage just yet. Chances are that with a little bit of work everything will be just as you intended. After all, you did create all this code and the hardware.

The problem is that there is nothing to look at or see; the thing is dead, and you don't know where to start. The solution is to make things visible and to start the process in a step-by-step manner so you can make sure each step in the program you have created is doing what it is supposed to do. The good news is that you do not have to spend a fortune on new software and hardware and you don't have to spend a year of your life learning a new discipline. You already know and have everything you need to debug the program in your LAB-X1 board. Even if the circuitry you are using is divorced from what is on the LAB-X1, the software can still be tested and we can still make sure the PIC is not dead.

Three output devices on the LAB-X1 board can be used as aids in the debugging process. They are

- The LCD
- The eight LEDs in the LED bar
- The piezoelectric speaker device

A number of input devices can also aid in the debugging process by making the debugging more interactive. These devices are

- The keypad
- The three potentiometers
- The reset button

Some standard software tools can also be used:

- The PAUSE command
- The STOP command
- The IF... THEN couplet
- The ability to COMMENT out sections of code

The PICBASIC PRO Compiler provides a number of statements that are designed specifically for the debugging process. These are mentioned earlier and should be studied in the PBP manual.

The personal computer you are using is also a powerful debugging tool in that it can both send and receive information and gives you a full screen and a keyboard to use as interactive elements. PICBASIC PRO provides a number of powerful tools to let you interact with your PC. However, the most powerful device at your command is *the computer between your ears*. By and large, the debugging process is an exercise in the use of your brain. Everything else necessary can be done with the LAB-X1 and your personal computer.

Following a few rules will make the debugging process both easier and more likely to succeed in a reasonable time.

Rule 1. Be thinking about the debugging process as you write the code. Design the code so it can be debugged, and put in the necessary hooks and connects as you go along.

Rule 2. Write the code as small subroutines that can be tested as stand-alone mini-programs. Once you have the software working, you can streamline the code. Test your program as you develop it to make sure each developmental level is operational.

Rule 3. Do not wait till the last moment to start the debugging process. Debug as you go along, meaning that you should debug the code as it is developed rather than waiting till it is all done and ready to be delivered to the customer. Learn to write the code so you can debug it as sections of code or as stand-alone subroutines.

Rule 4. Write a set of routines that can be called from within the code that shows you the content of various memory locations on the LCD or bargraph as the program runs.

The first thing most programs must do is make the LCD come alive. Two things must be checked to confirm its proper operation. Is the software right? Have you somehow destroyed the electronics?

SOFTWARE

PBP (PICBASIC PRO) makes it completely painless to use the LCD. All you have to do is tell the software where the LCD is connected to the hardware and which pins are connected to what function on the LCD. In the case of the LAB-X1 board we are using, the LCD is connected as follows:

The LCD is connected to PORTD and PORTE

It can use the 4-bit or 8-bit mode to send data to the LCD (if all pins are connected).

The register select bit is at PORTE bit 0

The enable bit is at   PORTE bit 1

The read/write bit is   PORTE bit 2 and is made low for writing to the LCD

These variables are defined by the following code segment. This code should be placed at or near the beginning of your program.

```
DEFINE LCD_DREG PORTD    ; LCD connected to PORTD
DEFINE LCD_DBIT 4        ; uses 4 bit data path
DEFINE LCD_RSREG PORTE   ; RESET register is PORTE
DEFINE LCD_RSBIT 0       ; uses bit 0
DEFINE LCD_EREG PORTE    ; ENABLE register is PORTE
DEFINE LCD_EBIT 1        ; uses bit 1
PORTE.2 = LOW            ; we will be writing to only LCD
```

You *must* have a pause of about 0.5 seconds at the start of your program, before you first access the LCD, to allow the LCD to complete its setup routines. If this pause is omitted, the LCD can malfunction or may not start up at all. Just to make sure, start with a PAUSE of 0.5 seconds and then shorten it when you know that everything is working properly.

```
PAUSE 500           ; and
ADCON1 = %00000111  ; set the digital modes needed. This has to
                    ; do with
                    ; making PORTE digital for controlling the LCD.
                    ; PORTE is analog on startup and reset. PORTD
                    ; is digital only and cannot be made analog.
                    ; Set TRISD.
```

This is needed because the 16F877A starts up and resets to analog mode. In analog mode all of PORTE and PORTA are in analog mode. We need the three PORTE pins to be in digital mode so we can control the LCD. The preceding instruction does this. It also makes PORTA digital, but that is not always necessary.

Go over your code to make sure all the preceding conditions are met *verbatim*, and are case perfect.

Write a short program to check the operation of the LCD hardware as comprehensively as you think is necessary. Program 23.1 can be used as a quickie starter.

Program 23.1 A short rudimentary program for testing the LCD

```
CLEAR                        ; clear variables
DEFINE OSC 4                 ; define osc
DEFINE LCD_DREG PORTD        ; define LCD connections
DEFINE LCD_DBIT 4            ;
DEFINE LCD_RSREG PORTE       ;
DEFINE LCD_RSBIT 0           ;
DEFINE LCD_EREG PORTE        ;
DEFINE LCD_EBIT 1            ;
LOW PORTE.2                  ; pull write bit low
                             ;
TRISD = %00000000            ; set all PORTD lines to outputs
TRISE = %00000000            ;
ADCON1 = %00000111           ; don't forget to set ADCON
                             ;
PAUSE 500                    ; pause .500 seconds for LCD startup
                             ;
LCDOUT $FE, 1                ; clear LCD, go to first line, first
                             ; position
LCDOUT "Now is the time for" ; print
LCDOUT $FE, $C0              ; go to second line
LCDOUT "a cup of pea soup!"  ; print
END                          ; end program properly
```

A similar but more comprehensive program that you wrote and that you are completely comfortable with should be in your utility files to let you check the proper hardware and software operation of your LCD whenever you think it is necessary. Your program should check every character and every command in the LCD's vocabulary if you want to perform a really comprehensive check.

Integer math is the source of a lot of problems for those who are unaware of the havoc that integer math calculations can visit on the software you are trying to debug. A certain amount of expertise with integer math is a must if you want to create mathematical routines within your software. If the routines are amenable to it, you should write a program around the routine to test every possibility that the routine might encounter and thus debug it the hard way even if it means your computer has to run the routine all night to get through all the commutations. Oftentimes, all that is necessary is to run the routines that would be called at the boundary conditions, or under the critical conditions, to make sure the routine is robust.

The following routine has a bug in it. See if you can find it. The routine is designed to make you aware of an 8-bit math problem.

We are trying to maintain a value of X = 127, but we need to do it in small steps. An external condition is changing X to anything from 1 to 255, and we cannot exceed these parameters because of the fact that X is a variable that has been defined as a byte.

```
MODIFYX:               ;
READ X                 ;
IF X<127 THEN X=X+5    ;
```

```
    IF X>127 THEN X=X-4    ;
    IF X<1 THEN X=1        ;
    IF X>255 THEN X=255    ;
    RETURN                 ;
```

Twos compliment convention is not supported in the integer math used by the PIC compiler, nor is implementation of the minus (–) sign. If you are having problems rewriting the math formulas with brackets, changing the order in which things are done might help.

Configuring the 16F877A and Related Notes

The LAB-X1 is set to 4 MHz by default. This is set by the ABC jumpers on the board. If A is set at 2–3, then it is 4 MHz. Since the compiler writes the default oscillator configuration to XT every time, 4 MHz causes the fewest problems for new users. This means the compiler always uses XT unless you inhibit it. You should inhibit this only if you are not running at 4 MHz. If you are running your own circuit, you will, in all probability, not be using a frequency dividing network (like the one the LAB-X1 uses), so your frequency will be the frequency of the crystal you use. The PIC adjusts the power used by the oscillator to minimize power use (critical for battery-powered devices). At the XT setting, the PIC puts out less power because at 4 MHz less power is needed. If you are using a 20-MHz crystal, you should use the HS setting for the oscillator. At the HS setting, the PIC provides more power for the oscillator. Some times the system will run a 20-MHz crystal at the XT setting, but the operation can be marginal.

When working on projects faster then 4 MHz, and where power consumption isn't critical, use a 20-MHz crystal. Uncheck the Update Configuration From File option in the programmer after changing to HS. This prevents the setting from reverting to XT every time a new hex file loads (as mentioned earlier).

The compiler always disables low-voltage programming by default. If you check the option Update Configuration From File in the programming software, this will make it to the chip. If you are setting configuration manually in the programmer software, simply select *Disable* for Low-Voltage Programming.

The configuration setting controls how the oscillator driver works on the PIC. It's mainly a question of power consumption, but there also seems to be some filtering involved. If you're using a 4-MHz crystal/resonator, the HS setting will work, but it will consume a bit more power than the XT setting. The XT setting won't drive the faster crystals reliably, presumably because it doesn't have the power.

The following is a situation for which there is currently no explanation. The LAB-X1 uses an external clock chip, which is essentially the same as a TTL oscillator source. When you use a *16F877* (not 16F877A) set to XT, it will work fine at 20 MHz. If you replace the PIC with a *16F877A*, however, it won't run at all. It requires the HS setting for 20 MHz, even though the oscillator source is external and the chip doesn't have to drive a crystal.

If pin B.3 is acting weird and the program randomly stops/starts/resets, it suggests that you have enabled Low Voltage Programming in the configuration screen of the programmer. You have to check this whenever you change the PIC you are using because it can change unexpectedly. (Meaning that I cannot predict when it will happen.)

QUESTION

If ADCON1 = 7, then does that mean that the TRIS registers have to be set for ports A and E to make them outputs. What is the effect of following ADCON1 = 7 with TRISA = 0 and PORTA = 0 on the PORTA A-to-D designations? This is important because the compiler does not clear the LCD on startup and can make it hard to see the effect of these commands in complicated circumstances.

ANSWER

The answer depends on what you are trying to accomplish. The TRIS registers are set to $FF on a reset, making all the pins inputs. If you need a pin on PORTA to be an output, you must clear the corresponding TRIS bit. Commands like LCDOUT set the TRIS automatically:

```
ADCON1 = 7   ; configures pins for digital operation
TRISA = 0    ; configures pins as outputs (PORTA value placed on
             ; pins)
PORTA = 0    ; changes PORTA value (drives all pins low)
```

If you want to insure that PORTA has a specific value when it becomes active as an output, set the TRISA value last:

```
ADCON1 = 7 ; configures pins for digital operation
PORTA = 0  ; changes PORTA value (pins still inputs - floating)
TRISA = 0  ; configures pins as outputs (PORTA value placed on pins)
```

The reset values of the registers are listed in the datasheets. They differ based on the type of reset, the PORT, and the PIC you are using. Some bits are set to 1 or 0, while others are specified as being "unknown" or "unchanged."

QUESTION

What if ADCON1 does not set all the A and E ports to digital? If a PORTA pin is still analog and we try to set it high with TRISA, what happens?

ANSWER

This also depends on the PIC and the pin in question. This answer will suffice in the majority of cases. The output circuitry is not affected by the pin being configured as analog mode. Therefore, you can write to the port as a whole and it will work as expected:

```
PORTA = $03
TRISA = $00   ; RA0 and RA1 go high, even if analog
```

The problems occur when reading the pins. A digital read of the port always returns zeros for pins configured as analog. Because of *the read-modify-write phenomenon, successive writes to different pins on a single port can produce unexpected results*:

```
TRISA = $00     ; set pins as outputs
PORTA.0 = 1     ; RA0 goes high
PORTA.1 = 1     ; RA1 goes high, but RA0 returns low unexpectedly
```

The behavior depends on the method used to send output. Avoid making changes one right after another to the same port to fix this! Add a short pause between commands.

The following applies to the Configuration and Option menus in the programmer software

Settings

Oscillator	XT (for 4-MHz crystal)
Watchdog	Enabled
Power up	Enabled (not critical)
Brownout	Enabled (not critical)
Low-voltage programming	*Must be disabled*
Flash program write	Enabled (critical only if using boot-loader)
No code or data protection	(Disable)

Set the oscillator to HS for speeds faster than 4 MHz.
Using a 20-MHz crystal is fine for most PICs. Check the datasheet for each unit.
Be sure to select the correct device number in the device ID box.

Configuration

Oscillator	4 MHz use XT
	20 MHz use HS
Code protection	Disable
Watch dog timer postscalar	Disabled
Brownout reset	Disabled

Enable	Watch dog timer enable	
	Power up timer enable	Disabled
Enable	Brownout reset enable	
	Low voltage programmer enable	Disabled
Enable	Flash program write enable	
	External data bus width	Ignore
	Mode	Ignore
	Memory size	Ignore

Options

Enable	Program/verify code	
Enable	Program/verify configuration	
Enable	Program/verify data	
	Program/verify ID	Ignore
	Program/verify oscillator calibration	Ignore
	Program serial number	Ignore
DISABLE	Update configuration	if not then 4 MHz
	Reread file before programming	Ignore
Enable	Erase before programming	
Enable	Verify after programming	
	18Fxxx file data address x2	Ignore
Enable	Disable completion messages	
	Skip blank check	Ignore

Simple Checks

The following are some simple checks you should apply if the PIC still does not oscillate.

Check that the power is on to the project.

Check that the power is on to the programmer.

Check that the programmer is plugged into your board.

Check that there is a device in the main PIC socket.

Check that the device in there is orientated for pin 1.

Check that the ZIP socket is locked in position.

Check that the correct device has been selected.

Check that the program has an END statement.

Check that the addresses for all the DEFINEs for the project are correct.

Check the spellings. (The compiler does not catch everything!)

Check that everything is in capital letters where required (see PBP manual).

Check that all sockets that should be empty are actually empty.

Run a test program "suited to and designed for" your specific project.

Use an oscilloscope to check that the system is actually oscillating at the OSC pins.

In most programs, some other pins should also be going high and low regularly.

Recheck the wiring.

Check for open wires and cold solder joints.

Check for short circuits.

Recheck the program.

Run a program that you know works on your PIC.

Some Programmer-Related Error Messages

If the programming connector is not connected, you will get a code check error.

If the programmer has no power, you will get a communication error.

If the wrong device is selected, you will get a blanking error.

Occasionally, if everything is okay, you will get a code check error. Just reprogram the chip.

Things I Have Noticed but Have Not Figured Out (and Other Mysteries)

If you know the answer, send me an e-mail so we can all share the information.

Sometimes, for some reason INTCON.0 will be set at 1 on startup. If you clear it, it does not reset itself.

On startup, OPTION_REG is %11111111, but if you set it to %01111111, the system hangs up if PORTB is not set appropriately to reflect the need for the Option Register setting. Meaning that if the PORTB pins are to be pulled up, some of them have to be programmed as inputs or the thing can hang up!

Assign the prescalar to the watchdog timer to get 1:1 pre-scaling on Timer2. See page 21 of the datasheet.

B7 and B6 (and sometimes B7 alone) are pulled low by the programming cable of the parallel programmer under certain conditions. This can inhibit the use of these two pins by the software while the programmer is connected to the board.

The global interrupt enable bit is cleared whenever an interrupt is set and reset automatically when the interrupt bit is cleared by the program. This means no new interrupt can be set till you clear the last interrupt.

Set only one interrupt at a time and follow through till it is cleared.

There can be more than one "ON INTERRUPT GOTO" call within a program, so each interrupt can be handled individually. It is best to stick with one interrupt till you become proficient at working with the PICs.

When using the LCD, the clearing routine at the beginning of the program must be something like:

```
PAUSE 500              ; pause for LCD start up
LCDOUT $FE, 1, "Clearing the LCD"    ; clear the display and
                                     ; show message.
PAUSE 250              ; This is useful for seeing a reset
                       ; button response
LCDOUT $FE, 1          ; clear again to make sure you are starting
                       ; with nothing in the LCD screen.
```

... because neither the compiler nor the CLEAR command clear the display on reset, and whatever happens to be in the display from the last program will stay in the display and mislead you. Since this can be confusing at the beginning of a program, it is your responsibility to clean this up.

Setting the Ports

All ports that will be used should be set for port configuration, and all pins that will not be used should be set as inputs. Setting them as inputs minimizes the possibility that an improperly set port pin will turn something on or off by mistake. It would not be out of line to actually set all other unused ports to inputs, though all ports are set to be inputs on startup. To know the exact official status of each port and register on startup and reset, see the datasheet. Some pins/bits may come up as undefined.

A cold start is not the same as a reset. If you are using the EEPROM in the chip, there can be added complications.

24

SOME REAL-WORLD PROJECTS YOU CAN BUILD

It is often hard to think up new projects that are within the scope of what we have just learned. The following are some useful instruments and controllers that you now have the skills to build. Each can be embellished to showcase your newly learned skills, and each one teaches you another skill. Some projects are harder than others. Points are awarded for how well the instrument works (engineering) and how good it looks (workmanship and design).

1. Build a finished useable instrument that will tell you how *fast you are accelerating or decelerating* as you move forward or backward in your car, and how many g-forces you experience as you go around a corner. Display the information in real time on the two lines of a 2-line-by-16-character LCD display. As an added bonus, you are asked to add an alarm that goes off if you exceed certain acceleration values and create the ability to record the maximum accelerations experienced during any one trip. (Would it be possible to integrate the g-force over time to tell how fast you were moving, and do you have the skills to do this? What are the problems involved and what kind of accuracy can you expect. Could some simple experiments allow you to determine if you are on the right track? What are these experiments?)
2. Build a rudimentary carpenter's level with a digital display that tells you how far from horizontal the level is in units of 0.1 degrees. Display your results on a 1-line-by-16-character LCD display. Added bonus: add an LED that blinks as long as the instrument is kept level within 0.3 degrees to the horizon.
3. Build a rudimentary *digital protractor* that will allow you to place an object at any angle to the horizon within 0.5 degrees. (Hint: You must use both axes of the Memsic and switch between them at 45 degrees.) Display your results on a 1-line-by-16-character LCD display. Added bonus: provide a written discussion of the accuracy that can be obtained with the detector being used, and discuss what the limiting factors are. Discuss what fabrication problems you encountered during construction

and what the software problems were. Where is the protractor most and least accurate? Provide a table of + and − error probability values for every 5 degrees.

4. Build an *18-inch diameter wall thermometer that uses one R/C servo* to directly connect to and drive the indicator/needle of the thermometer. These servos have a range of about 180 degrees. Use 100 degrees of this range to indicate a temperature from −30 to +120 degrees. Use the LM34 as your temperature sensor.

5. Build a *12-inch diameter thermometer similar to the one in project 4 that uses a small and inexpensive 200-step-per-revolution stepper motor* as the driver for the thermometer needle. Use an appropriate thermistor as your detector and calibrate it for this application. Since the position will be lost during a power failure, your software must provide the features needed to zero against a stop and thus start the servo from a known position every time. This is the technique used for most computer-controlled instruments in automobiles and in industry. (All this can be done with cardboard construction techniques.)

6. Build a small voltage monitor to *monitor and display the conditions of your car battery terminals* at all times. Design this so it can be mounted on your dash.

7. Build an accessory for your soldering iron that lets you *control the temperature at the tip of the iron* by turning the iron on and off. Display the tip temperature and provide an LED to indicate the "power on" at the iron. Use a tiny thermistor attached a $1/2$ inch from the tip as your sensor. The temperature should be nearly constant across the copper tip.

8. Build a combination *speedometer and odometer* for use on a bicycle. The unit must detect wheel rotation at least four times per revolution. Allow the input of the wheel diameter in inches.

9. Build a detector and motor driver that will allow a *solar collector to remain aimed at the sun all day* to optimize the collection of solar energy. Power it using a stepper motor or a servo motor. Only one axis needs to be moved.

10. Build a $1/4$-*second pendulum that is accurate to one cycle of the MCU clock*. An iron bob on the end of a string will do. Provide the software to slow down or speed up the pendulum as necessary to stay in time with the signals provided by the microprocessor clock, the $1/4$-second time signal that you will have created for this project. (Hint: You will need a couple of coils to push and pull on the iron bob to keep the system in sync with the microprocessor signals.) This exercise should demonstrate that even fairly simple techniques can be used to create an accurate pendulum for a clock. Discuss how this technique could be used to improve the accuracy of an aging grandfather clock.

11. Design, build, and patent a device that will indicate the *oil level in an automobile* with a ten-segment bargraph. The top two LEDs will indicate overfill. The two below them are to indicate perfect oil level. The next four will indicate acceptable oil level with each one indicating, relatively, how much oil is in the pan. The last two should indicate that oil is too low and thus not allow you to start the vehicle.

12. Design and build a marketable version on the *dual thermometer* device that can be placed across your hot air furnace or boiler to constantly indicate the temperature difference being generated by the heat/cooling device. Both cooling and heating seasons are to be accommodated with automatic selection of the season by the device. Provide a complete set of instructions for use by a nontechnical customer.

13. Design and build the *logic works for a large clockwork about 4 feet in diameter*. The system is to put out a square wave TTL-level signal every 0.1 seconds. This signal will be employed by the user to feed into a clockwork driven by a stepper motor and suitable drive reduction belts and gears. The power to the stepper motors is to be provided by others. Provide a readout at the device that displays the time on an LCD display, and also provide inputs for six buttons to allow the time to be adjusted, up and down, with two buttons each for hours, minutes, and seconds. Provide an interlock switch. In the locked position, the switch will interlock the display to the 0.1-inch signal. In the unlocked position, the signal will go out (clock hands will move) but will be isolated from the LCD display. This will allow the clock to be synchronized to the display after startup, or if the power goes down, or if some other malfunction occurs, offering a way to move the clock hands on the control panel. This means that if you ask the hands to move forward 5 minutes, the controller has to put out enough pulses at a rapid rate to make this move. See the next project.

14. Design and *build the amplifier needed to run the stepper motors* in Project 13. Amplifier shall provide signals to the stepper motor coils and provide all the connections needed to connect to the stepper motor. Provide a way to divide the input signal frequency in case the gearing on the clockworks does not match the 0.1-second pulses from the logic works. This is to be done in the electronics within the amplifier module. Add convenience hardware and software features as you see fit to create an easy-to-use marketable amplifier module.

Finally, describe in detail five projects you think you could build using what you have learned. File your notes in your shop manual for future reference.

CONCLUSION

After it's all said and done, making an instrument or a controller is a matter of putting together a series of components and segments of programs, each of which provides a specific function not unlike what we have done in almost all the projects in this tutorial.

In this tutorial, we have covered the most basic techniques for doing this. Other more sophisticated techniques should not be any harder to investigate and assemble. Writing a short program to investigate what needs to be done for any part of your project should not be difficult with the expertise you now have. Incorporating the code into a larger program can get complicated if timer constraints get in the way. Usually, the most difficult task will be getting the programs to run fast enough to get the job done in the time available. We saw this in the programs that had to update the seven-segment displays. However, we were running all the programs at 4 MHz and you will find that considerably more can be done at 20 MHz. Most of the PICs can be run at 20 MHz, while some of the newer ones can be run at 40 MHz.

Repeated calculations and comparisons are time consuming, as is writing to the LCD. Iterations can take up a lot of time and should be avoided. Avoid or at least reduce the number of times a calculation is performed and the LCD is written to. If a calculation can be done up-front, do it and store the result. Avoid doing the same calculation over and over again in a loop.

It is well worth your while to learn how to really use the timers and counters. They are the key to getting a lot of things done right, fast, and with the proper timing. Build your programs up a line of code at a time and be sure you understand exactly what each line of code does.

The 16F877A has 8K of memory. Most of the programs we wrote were in the 400- to 800-word range, so considerably more sophisticated programs can be written without adding any memory. On the other hand, adding one-wire memory is neither hard to do nor expensive.

Only a few instructions are used in the projects we undertook. This was done to keep the emphasis on the development of the projects as opposed to learning what wonderful tricks could be done with the language and discovering how powerful the language was.

Expanding the number of instructions you are comfortable with will make your projects more powerful. The first half of the book will help you in this direction.

Circuit diagrams are provided for all the projects in order to help you get comfortable with designing your own projects. As you can see, this is not overly difficult. All the drawings I made are on the support Web site in AutoCAD format and you can cut and paste from them to speed up your work. Each is also available in the Adobe .eps format there, but they are saved as individual files and are much harder to use than the AutoCAD file.

Oftentimes, it might be necessary to use more than one interrupt and have more than one timer or counter in operation. This can get complicated, and I gave no hint on how to proceed when this is the case. At some 300 pages, this tutorial is already getting too long. Those techniques will have to wait for the next, more advanced, text.

APPENDIXES

SETTING UP A COMPILER FOR ONE-KEYSTROKE OPERATION

It is possible to set up the microEngineering Labs programmers for one-keystroke programming so that one keystroke (F10) or a mouse click on the Compile and Program icon will.

1. Open the programmer software
2. Compile the program
3. Check it for errors
4. Send the program to the PIC in the programmer
5. Shut down and close the programmer window
6. Run the program in the PIC

The following instructions outline the process:

1. Open the MicroCode Studio editor
2. Pull down the View menu and select PICBASIC options

   Under the programmer bar, select the programmer you are using as the default programmer.
   Click Edit and enter the name of the programmer file (meprog.exe).
   Click Next.
   Let the program find the file. If it cannot find the file for it manually...
   Click Next.
   In the file name parameters, at the end of the line add a space, -p, another space, and -x.
   The entry now ends in... filename -p -x
   Do not omit the two spaces preceding each - sign.

   Having done this, whenever you want to transfer the program you are editing to the PIC, just press F10 or click the compile and program icon.

B

ABBREVIATIONS USED IN THIS BOOK AND IN THE DATASHEETS

A dictionary of PIC-related terms for the uninitiated.

A-to-D	Analog-to-digital
ADC_BITS	Analog-to-digital converter. Sets the number of bits that the analog-to-digital conversion uses, usually 8 or 10.
ADC_CLOCK	Analog-to-digital converter. Defines where the clock or the process will be read from.
ADC_SAMPLEUS	Analog-to-digital converter. Defines the sample rate in micro (US is for micro) seconds.
ADCIN	Analog-to-digital converter. Instruction to read the analog-to-digital channel input line selection; the identification number that follows tells which line to read.
BASIC	An easy-to-learn language for programming computers.
BIT	A 1-bit variable. Can hold a 0 or a 1.
BOR	Brown out reset
BYTE	An 8-bit variable. Can hold a number from 0 to 255.
CLKIN	Clock input line
CMOS	Complementary metal oxide semiconductor
CPU	Central processing unit
CS	Chip select
DC	Direct current

EEPROM	Electrically erasable programmable read-only memory
EPIC	Programmers made by microEngineering Labs
EPROM	Erasable programmable read-only memory
FLASH	Memory that can be programmed electrically in a flash
I2C	I two C, a type of serial memory
IC	Integrated circuit
ICSP	In circuit serial programming
IR	Infrared
LAB-X1	Experimental board's name
LCD	Liquid-crystal display
LCD_DBIT	Liquid-crystal display. Data bit
LCD_DREG	Liquid-crystal display. Data register
LCD_EBIT	Liquid-crystal display. Enable bit
LCD_EREG	Liquid-crystal display. Enable register
LCD_RSBIT	Liquid-crystal display. Register select bit
LCD_RSREG	Liquid-crystal display. Register select register
LCDOUT	Liquid-crystal display output. Sends the information to the liquid crystal.
LED	Light emitting diode
mA	Milli amps – 0.001 amps – 10 to –3rd
uA	Micro amps – 0.000001 amps – 10 to –6th
MCLR	Master clear—resets the chip on startup
MCU	Micro controller unit; the PIC 16F877A is a microcontroller.
MHz	Mega hertz; 1,000,000 cycles per second
NIBBLE	Half a byte. A 4-bit variable. Can hold a number from 0 to 16.
Ns	Nanosecond; 0.000,000,001 seconds; 10 to –9th.
PCB	Printed circuit board
PIC	Peripheral interface controller; original name for microcontrollers
POR	Power on reset
PRO	Professional

ABBREVIATIONS USED IN THIS BOOK AND IN THE DATASHEETS

PROM	Programmable read-only memory
PSP	Parallel slave port
PWM	Pulse width modulation
PWRT	Power up on timer
RAM	Random access memory
RC	Resistor-capacitor usually used with oscillators
RISC	Reduced instruction set computer
RS232	A communications standard for short-haul communications
RS485	A communications standard for longer haul communications
SPI	A one-wire standard
SSP	Synchronous serial port
SST	Oscillator startup timer
TOCKI	Timer zero clock 1
TRISA	Tri state register "A"
VAR	A declared variable. All variables used must be declared up-front in PICBASIC PRO before their use.
Vcc	Power for the devices—for example, for the motor you are running.
Vdd	Logic power voltage level for chips as regards supply power
Vss	Logic ground for chips as regards logic supply power
WDT	Watchdog timer
WORD	A 16-bit variable. Can hold a number from 0 to 65,536.
ZIF	Zero insertion force (socket)

C

LISTINGS OF PICBASIC PRO PROGRAMS ON THE INTERNET AT MELABS.COM

All the programs that apply to this discussion, written by microEngineering Labs, are also available on the support Web site in one searchable file.

Having all the programs in one file allows you to search for any one word, command, or structure in all the programs to see how it was used in this book. You can also cut and paste sections of code from these programs to your programs.

This is a file that lets you copy the programs from the support Web site and run them on your computer. It saves you the time of having to retype the programs.

D

NOTES ON DESIGNING A SIMPLE BATTERY MONITOR INSTRUMENT: THINKING ABOUT A SIMPLE PROBLEM OUT LOUD

This appendix is about how to think about a problem.

When you decide you need to design an instrument or device to perform a specific task, the first thing you need to know is exactly what you want the device to do. This is not as simple as it seems at first glance. Let's take a close look at a fairly straightforward example.

We have a small lead cell battery that provides about 6 amps hours at about 12 volts. We need to monitor the condition of this battery at all times for a relatively critical application. The simple answer to this is that we need to "monitor the voltage of the battery" to see if everything is OK. That is, of course, true, but there is a lot more to it than that. Let's let our imagination run wild for a moment and think about what might be possible with the skills we are acquiring.

We could say: all we need to see is a green LED for "All is OK" and that is it!

But we can do a lot more than that when we become expert MCU programmers. Let's take look at what else might be possible.

If the situation is intermediate, we can turn on a yellow LED.

If things are getting near needing attention, we can turn on an orange LED.

If things are bad, we can turn on a blinking red LED.

We can use a test button to put a small load on the battery for one second and then report the voltage on one line of an LCD.

We could reset a timer that tells us when the battery was last tested and display the time elapsed on the LCD.

We could use a real-time clock function that tells how long it has been in days, hours, minutes, and seconds since the last full charge. We want this to reset automatically when the battery is charged. We can display this on the second line of an LCD.

We might want to display how many volts the charge got the battery up to, to gauge the condition of the battery as it ages. We might want to wait 15 minutes before taking the reading to let the cell settle down first so as to make sure our reading was more accurate.

We could monitor the voltage of the battery once a day and then turn on the charger for a specific time to recharge the battery so we do not overcharge the battery and dry it out.

We could monitor each cell in the battery separately and report on its condition constantly if we were in a space capsule.

We could transmit the condition of the battery to a central maintenance location once a day for maintenance follow-up.

We could add a sleep cycle to our program to minimize our instrument's current draw from the battery.

In order to affect most of the preceding points, we will need a certain amount of hardware. Let's take a closer look at what we need:

An LCD display with two lines of 16 characters each

Four LEDs: green, yellow, orange, and red for our indicators

One MCU with the following minimum properties:
    Seven lines for the LCD
    Four lines for the LEDs
    One output line for the charge connection switch
    One input line to measure the voltage
    One input line to reset and/or start measurements
    Two lines for a communications port, RS232 protocol

All this is well within the capabilities of the PIC 16F877A we have been experimenting with, but there are a number of less expensive MCUs that can do the job.
How could we improve on this design?
Since you may well be using a small lead acid battery in your shop (it is often the best of all possible choices for a rechargeable shop battery), you may want to implement a controller for your own personal use based on the previous discussion.

E

USING THE SUPPORT WEB SITE TO HELP MAKE INSTRUMENTS AND CONTROLLERS

Most of the information you will need has been provided on the Web sites* that supports this book.

Brief Description of All Files on the Internet and Their Intended Uses

Text-only listing of all microEngineeing Labs program files.
    Having all the programs in one file lets you search for any command you are interested in to see how it was used in a program by microEngineering Labs.

Text-only listing of individual programs by section in this book.
    This listing allows you to copy programs and run them. These programs may not be exactly what is in this book for the LAB-X1. These are the latest versions.

Text-only listing of all the programs in this book in one file.
    Having all the programs in one file allows you to search for any command you are interested in to see how it was used in a program by the author of this book.

Color photographs of various items.
    Reference material.

Drawings for the solar collector.
    Drawing with dimensions. AutoCAD file.

*Internet support sites: www.encodergeek.com and www.mhprofessional.com/sandhu.

Schematic with all the files in it.
> Schematic file with all the circuit diagrams from this book in it (in AutoCAD file format). For use in creating your own diagrams and designs. AutoCAD file.

Terminal setup.
> Terminal programs.

Book reviews.
> Short reviews of books related to PIC microprocessors.
> Helps you decide on certain very basic level book purchases.

INDEX

2-line-by-16 character LCD module, *142f*
3-volt signals, *171f*
4 MHz setting, 324
5-volt TTL-level signals, *170f*
6-inch rulers, 161
12-volt signals, 169–171
15-volt logic, *170f*
24-volt signals, 169
40-pin MCUs, 8
40-pin PICs, 11

abbreviations, 341–343
ac (alternating current), 166
acceleration instruments, 331
ADCON0 register, 75
ADCON1 register
  A-to-D conversions in, 75
  controlling digital and analog settings, 56
  dual thermometer instruments, 264
  setting values before LCD use, 44
  using PORTA as digital device, 29–30
aircraft servomotor control, 238
alternating current (ac), 166
amplifiers, 333
analog devices, 29
analog input pins, 259–261, *263t*

analog-to-digital conversion capabilities, 75, 140
PIC 16F877A microcontroller unit
  capacitance, 28
  frequency, 28
  overview, 26–27
  POT command, 27–28
  reading switches, 28
  voltage, 28
artificial horizon table project
  building, 275–277
  discussion, 270–271
  gravity sensor exercises, 277
  hardware connections
    overview, 271–273
    single-axis software, 273–275
    two-axis software, 275
  overview, 181, 269–270
Axelson, Jan, 115

Bargraphs exercise, 82
BASIC Compiler instruction set, 35–37
batteries, 166–167, 308
battery backup, clock, 123
battery monitor instrument, 347–348
battery terminal monitors, 332
beeps, 60–63

bicycle instruments, 332
binary notation, 29, 44, 51
binary values, 56–57, 74–76
biological sensors, 160
Bit C0, 31
Bit C1, 31–32
Bit C2, 31
Bit C3, 31
Bit C4, 31
Bit C5, 31
Bit C6, 31
Bit C7, 31
blinking LEDs, 43–44
boards, microEngineering Labs, Inc., 16
Bray terminal program, 319
breadboarding, PIC 16F877A microcontroller unit, 8–9
busy flag, 151–154
Busy Flag/Address Read command, *156t*
bytes, Timer1, 224

CA (common anode) displays, 195–197
capacitance, 28
car battery terminal monitors, 332
car odometers, 161
car speedometers, 160
cardboard table mechanism, 276–277

353

CCP trigger output, 257
CD-ROMs, 18
circuitry diagrams
    additions to tachometer for metronome, *225f*
    for controlling an RC servo from potentiometers, *65f*
    dual thermometer, *262f*
    generating tones on piezo speaker, *60f*
    input from Hall effect sensor into PICs, *193f*
    input of tachometer signal into PICs, *193f*
    LED bargraph, *49f, 51f*
    overview, 45
    potentiometer, *59f*
    pulse generator, *190f*
    thermistor connection to PIC, *297f*
Clear Display command, *155t*
clock frequency, 83
clock ICs, 122–124
clocks, 306, 333
codes, LCD, *141t*
common anode (CA) displays, 195–197
compilers, 17
    BASIC Compiler instruction set, 35–37
    PICBASIC PRO Compiler (PBP)
        example program, 43–44
        free demo, 45
        overview, 42–43
        tips and cautions, 44–45
    PICBASIC PRO Compiler instruction set
        math functions/operators, 40–42
        overview, 37–40
        setting up for one-keystroke operation, 339
conditioning signal
    input
        alternating current (ac), 166
        direct current (dc), 166–167, 169–171
        keyboards, 167
        overview, 165
        relays, 167–168
        resistances, 168

conditioning signal (*Cont.*):
    output
        inductive loads, 174
        parallel interface, 174
        resistive loads, 174–175
        serial interface, 173
connections, programmer, 221, 266, 271
control codes, 80, 144
cooling fans, 301, 303
cost of components, 16
COUNT command, 190–192
counters, 34
    exercises, 112
    function of, 104–105
    overview, 83, 104–111
    prescalars and postscalars, 111
    using Timer0 as, 105–107
    using Timer1 as, 107–111
crystal oscillation
    checks on hardware side, 316–317
    checks on software side, 317–318
crystals, 121–123, 316
Cursor at Home command, *155t*
Cursor/Display Shift command, *155t*

data logging, 179
data sheets
    abbreviations, 341–343
    PIC 16F877A microcontroller unit
        downloading, 10–11
        fast Internet connections, 10
        overview, 9–10
dc (direct current), 166–171
dc motors, *189f*
de-bouncing capabilities, 28
debugging
    commands that provide debug output to serial port, 319
    configuration, 326–327
    configuring 16F877A, 324–326
    crystal oscillation
        checks on hardware side, 316–317

debugging (*Cont.*):
    checks on software side, 317–318
    dumb terminal programs, 319
    if chip refuses to run, 318
    options, 327
    overview, 315
    at practical level, 320–324
    programmer-related error messages, 328
    setting ports, 329
    settings, 326
    simple checks, 327–328
    solderless breadboards, 320
    using PBP Compiler commands, 319
deceleration instruments, 331
decimal values, 56–57, 74–76
decision-making capability of instruments, 183–184
decision-making code segment, 204
DEFINE statements, 44–45, 53, 55, 139–140
devices, LAB-X1, 6
diagrams
    circuitry
        additions to tachometer for metronome, *225f*
        for controlling an RC servo from potentiometers, *65f*
        dual thermometer, *262f*
        generating tones on piezo speaker, *60f*
        input from Hall effect sensor into PICs, *193f*
        input of tachometer signal into PICs, *193f*
        LED bargraph, *49f, 51f*
        overview, 45
        potentiometer, *59f*
        pulse generator, *190f*
        thermistor connection to PIC, *297f*
    wiring
        counting into counter TMR1, *249f*
        IR-LED phototransistor pairs, *240f*

diagrams (*Cont.*):
  programmable pulse counter, *201f*
  set of 4 seven-segment displays, *197f*
  seven-segment displays, *199–200f*
  single IR transmitter–phototransistor pairs, *284–285f*
  single-point programmable controllers, *295f*
  solar collector instrument, *305f*
  testing sensitivity of phototransistor response, *254f*
  testing TMR1 counter, *253f*
  touch panels, *282f*
digital devices, 29
digital mode, 55
digital protractors, 331–332
diodes, 174, 317
direct current (dc)
  circuitry for conditioning signals, 169–171
  overview, 166–167
disabling low-voltage programming, 324
Display ON/OFF Control command, *155t*
DOS environment, 15
downloading data sheets, 10–11
DS1302 real-time clock, 123–124
DS1620 temperature sensors, 128–129
DS1820 temperature reading devices, 126–128
dual thermometer project, 180–181, 259–267
dual-axis artificial horizon program, 275
dumb terminal programs, 132–133, 319

editing software, 5
Editor exercise, 82
EEPROMs, 115
electrical signals, 161, 165
encoder pulse counter program, 106–107
Entry Mode Set command, *155t*

Epson SED series controller, 143
errors, syntax, 17
escapement, counting with, 240–243
escapements, 234–237
Evil Genius books, 164
execution routines, 286
expansion, PIC 16F877A microcontroller unit, 7–9
external interrupts, 30

fans
  cooling, 301, 303
  solar data collector project, 310
fast Internet connections, 10
feedback devices, 318
folder, PIC Tools, 16–17
Forty Characters exercise, 82
Fosc/4 frequency, 83
Four lines exercise, 82
frames, touch panel, *280f*, 281
FREQOUT command, 63
frequency, 28, 166
Function Set command, *156t*

gates, escapement, 237
global interrupt enable (GIE), 213–214
gravity sensor exercises, 277
gravity sensors, 269–270, 272–273

hall effect sensors, 192, *193f*
hardware programmers, 5
hardware pulse width modulation (HPWM) command, 61–62, 104, 212
hardware setup, 13–15
hearing, 159
"Hello World" program, 53–55
HEX files, 17–18
hex values, 56–57, 74–76
high nibble, PORTB, 71
high voltage, 166
Hitachi 44780 datasheet, 141
Hitachi HD44780U controller, 143
"hot air" solar collectors, 302
house thermostats, 160
HPWM (hardware pulse width modulation) command, 61–62, 104, 212

HS setting, 324
HSEROUT command, 134
HyperTerminal program, 133, 136

I2C memory, 115–117
I2C SEEPROM memory, 113
IC DS1202 clock, 122–123
IC DS1302 clock, 122–123
IC NJU6355 clock, *121f*, 122–123
inductive loads, 174
infrared (IR) light beams, 281, 283–284, 287
inhibit code, 296
initialization, LCD, 149–151
input capabilities, LAB-X1, 5
inputs/outputs
  beeps, 60–63
  creating, 48–50
  exercises
    controlling LCD, 80–81
    controlling LEDs, 80
    miscellaneous, 81–82
    overview, 79
  flexibility, 78–79
  LCD displays
    digital and analog settings, 56
    overview, 53–55
    reading keyboard and displaying key number on, 73–74
    writing binary, hex, and decimal values to, 56–57
  LEDs
    blinking eight in sequence, 51–52
    blinking one, 50–51
    dimming and brightening one, 52
    overview, 47
  potentiometers
    reading and displaying results on LED bargraphs, 57–59
    reading and displaying value on LCD, 74–76
    reading three and displaying values on LCD, 76–77
  programs that create inputs/outputs, 48

inputs/outputs (*Cont.*):
   RC servos, controlling from keyboard, 63–67
   reading inputs, 67–72
      reading keyboard, 67–70
      reading keyboard and displaying value on LCD, 70–72
      turning on LED only while button is down, 67
INTCON (interrupt control register)
   bit values of, 89–90
   bits 6 and 7, 103
   bits that affect Timer0, 215
   marble counter project, 258
   PORTB interrupt capability, 30–31
   Timer1 control, 107
integer math, 317, 323–324
interactivity of instruments, 183
interrupt capability, PORTB, 30
interrupt control register. *See* INTCON
interrupt enable bits, 213
interrupt flags, 213, 249
interrupt latency, 213
interrupts, 212–213. *See also* timers
   counting to register using, 244–246
   handling routines, 84, 213
   intervals between, 98–99
   lost by PAUSE commands, 83
   routine for Timer0 program, 216–217
   structure of routine of, *99f*
   timers, 33–34
intervals, timer, 98–99
"IntRoutine" routine, 88
I/O interfaces, LAB-X1, 5
I/O pins, 10
IR (infrared) light beams, 281, 283–284, 287
IR-LED phototransistor pairs wiring diagram, *240f*
isolators, 170

J7 Servo position control, using servo at, 66–67
jumper J4, 123
jumper J5, 121

keyboards, conditioning output signal, 167

LAB-X1 board
   debugging with, 320–321
   determining use of pins and ports, 23–26
   hardware features, 5–6
   image of, *27f*
   metronome program based on Timer0 and, 220–221
   modifications for solar collector, 304
   overview, 4
   programmable tachometer project on, 187–188
LCD (liquid crystal display)
   ADCON1 settings, 44
   code table, *155–156t*
   codes, *141t*
   connecting, 322
   design intent
      busy flag, 151–154
      hardware needed, 146
      information needed, 147–151
      programmer needed, 146
      software needed, 146
   digital and analog settings, 56
   exercise, 80–81
   exercises, 154–156
   hardware and software interaction, 143–144
   hardware of, 144–146
   module, *142f*
   overview, 53–55, 139–142
   reading keyboard and displaying key number on, 73–74
   solar data collector project
      overview, 310–311
      problem, 314
      sending data to computer, 311–314
   talking to, 144
   testing program, 323
   in touch panel programming, 286–287
   using in projects, 142–143
   writing binary, hex, and decimal values to, 56–57
lead acid batteries, 166–167

LEDs (light emitting diodes)
   blinking eight in sequence, 51–52
   blinking, EX, 43–44
   blinking one, 50–51
   dimming and brightening one, 52
   dimming using PWM command, 61–62
   exercise, 80
   interrupt-driven tasks, 85–88
   program using Timer0, 101–102
   seven-segment displays, 195
   in temperature controlling device, 185
   in touch panels, 279, 281
   turning ON, 67
   turning ON and OFF program, 49
light emitting diodes. *See* LEDs
liquid crystal display. *See* LCD
LM34 temperature sensors, 259
LM34-based controller, 296
loading software, 15
loops, 318
low nibble, PORTB, 71
low voltage, 166
low-voltage programming, 324–325
LTC1298 12-Bit A-to-D converter, 124–126

magnitude, dc signal, 167
marble counter project, 178
   counting
      directly into internal counter, 246–248
      with escapement, 240–243
      to register using interrupts, 244–246
   notes, 243
   overview, 180, 233–238
   servos, 238–239
   Timer1
      in asynchronous counter mode, 256–257
      in counter mode, 248–255
      INTCON interaction, 258
      oscillator, 257
      prescalars, 257

marble counter project (*Cont.*):
  resetting register pair (TMR1H, TMR1L), 257
  resetting using CCP trigger output, 257
  Timer2, 258
materials, solar collector, *304t*
math operations
  BASIC Compiler instruction set, 37
  PICBASIC PRO Compiler instruction set, 40–42
MCLR pins, 316
mechanical stop watches, 161
memory, non-volatile, 314
Memsic 2125 dual-axis accelerometer, 269–270
metronome project
  overview, 179–180, 209–211
  terminology, 212–214
  Timer0, 211, 214–224
  Timer1, 211, 224–227
  Timer2, 212, 228–230
  watchdog timer, 230–231
Microchip MPLAB editor, 42
Microchip Technology Corporation Inc.
  16F877A Microcontroller datasheets downloads, 23
  overview, 21
  Web site, 10–11
MicroCode Studio editor
  checking for problems with, 317
  "one mouse click", 339
  overview, 41–42
  program shortcut, 17
  starting programmer software from, 14
microcontrollers, 316
microEngineering Labs
  notes on software, 18–19
  overview, 21
  PIC 16F84A program, 153, *154f*
  preassembled boards, 16
  program for DS1620 readings, 128–129
  program for DS1820 readings, 126–128
  program to read from 12-Bit LTC1298 A2D chip, 124–126
  RS232 Communications program, 131–132

microEngineering Labs (*Cont.*):
  voucher for, 16
  Web site, 4, 10, 45
Microwire memory, 115, 118–120
Microwire SEEPROMs, 118–120
Mirror exercise, 82
Model ACM 1602K display, 147
Motorola Semiconductor, 115
motors, dc, *189f*

National Semiconductor, 115
National Semiconductor LM34 temperature sensors, 259
NMBR, POT command, 28
nonvolatile memory, 314

odometers, 161, 332
oil level reading devices, 332
ON INTERRUPT, 90–93, 258
"one mouse click" set up, 43
open drain output, 30
OPTION_REG (option register), 88–89
  bit 6, 30–31
  bit assignment, 88–89
  bits that affect Timer0, 214–215
  effect on Timer0 counter, 107
  using Timer0 as counter, 104
opto isolators, 170
oral thermometers, 160
oscillators
  marble counter project, 257
  speed, 44, 83, 132
oscilloscopes, 316
output capabilities, LAB-XI, 5
output information, instrument's, 184
output set point control devices, 185
oven thermometers, 160

Parallax Inc., 269
parallel interface, conditioning input signal, 174
parallel port programmer, 14
PAUSE commands, 83, 213
PAUSEUS command, 274
PBP. *See* PICBASIC PRO Compiler
PC boards, 259
pendulums, 332

period register, Timer2, 102
phototransistors, *254f*, 283–284, 287
PIC 16F84A program, 9, 153, *154f*
PIC 16F877A microcontroller unit
  40-pin MCUs, 8
  40-pin PICs, 11
  additional hardware, 7–8
  breadboarding, 8–9
  core features, 22
  data sheets
    downloading, 10–11
    fast Internet connections, 10
    overview, 9–10
  expansion, 7–9
  overview, 3–6, 21–22
  peripheral features
    analog-to-digital conversion capabilities, 26–28
    overview, 23–26
  ports
    configuring and controlling properties of, 29
    PORTA, 29–30
    PORTB, 30–31
    PORTC, 31–32
    PORTD, 32–33
    PORTE, 33
  precautions, 9
  software compiler, 7
  timers
    counters, 34
    overview, 33–34
    prescalars and postscalars, 34
    watchdog, 34
PIC Tools folder, 16–17
PICBASIC PRO Compiler (PBP), 4, 7, 15
  commands, 319
  example program, 43–44
  free demo, 45
  instruction set, 37–42
  overview, 42–43
  tips and cautions, 44–45
PICBASIC PRO programs, Web site, 345
PIE1 control register, 108
pin B7, 9
pin out designations, 40-pin 16F877A PIC microcontroller, *8f*

Pin, POT command, 28
pins
    analog input, 259–261, *263t*
    LAB-X1 board, 23–26
    LCD, *145t, 147–148t*
PIR1 control register, 107
PLCs (programmable logic controllers), 243
PORTA, 29–30
PORTB, 30–31
    lines set as inputs, 26
    reading keyboard, 70–71
    reading switches, 67–68
    Servo Position Control for an R/C servo from, 63–64
PORTC, 31–32
PORTD, 32–33
    LCD data, 53
    loading potentiometer readings, 58
PORTD LED bargraph circuitry, *51f*
PORTE, 32, 33, 53
ports
    configuring, 29
    controlling properties of, 29
    LAB-X1, 23–26
    PORTA, 29–30
    PORTB, 30–31
    PORTC, 31–32
    PORTD, 32–33
    PORTE, 33
    setting, 329
postscalars, 34, 85, 99, 103, 111, 212
POT command, 27–28
potentiometers
    controlling an RC servo from, 64–65, 78–79
    reading and displaying results on LED bargraphs, 57–59
    reading and displaying value on LCD, 74–76
    reading three and displaying values on LCD, 76–77
    in single set point controllers, 294
    for single-axis artificial horizon, 274
    in solar collector, 304
    in temperature controlling device, 185
power-up, LCD, 149–150, 322

prescalars, 34, 212
    counters, 111
    increasing time between interrupts, 99–100
    marble counter project, 258
    overview, 85
    program to see effect on Timer0 operation, 218–219
    Timer1, 100
    for Timer2, 103
    on watchdog timers, 104, 231
programmable logic controllers (PLCs), 243
programmer-related error messages, 328
programmers, 5, 14–15, 146
projects
    artificial horizon table, 181
    dual thermometer, 180–181
    marble counter, 180
    metronome, 179–180
    single-point controller, 181
    solar collector, 182
    tachometer project, 179
    techniques, 177–182
    touch screen, 181
    Web site, 177
Proton+ Lite BASIC editor, 42
protractors, digital, 331–332
proximity detectors, 243
pulse generators, 189, *190f, 191f*
Pulse Width Modulation (PWM) command, 60–62, 104

quotation marks, 44

RadioShack, 164
R/C servos, 78–79
RC servos, controlling from keyboard, 63–67
Read Data command, *156t*
reading clock program, 309
reading switches, 28
real-world projects, 331–333
receive buffers, 135
register names, 44
relays, 167–168
resistance, 26–28, 57–58, 168, 298
resistive loads, 174–175
resistor R17, 9

RS232 Communications program, 131–132, 134–136
RS485 communications, 137
rulers, 161

saving programs, 18
Scale, POT command, 28
scaling ability, timer. *See* postscalars; prescalars
scan routines, 286
Schmitt trigger input buffers, 31–33
sensors (transducers)
    overview, 159–161
    reasons for building, 161–162
    resources, 164
    types of, 163–164
serial EEPROMs, LAB-X1, 6
serial interface, conditioning input signal, 173
serial one-wire memory devices, 113–115
serial peripheral interface (SPI) memory, 115, 117–118
servos, 238–239
Set CG RAM Address command, *156t*
Set DD RAM Address command, *156t*
set point adjustment devices, 294
seven-segment displays, 195–207
shortcut, MicroCode Studio program, 17
sight, 159
signal latching and clearing, *168f*
single-axis software, 273–275
single-point controller project
    control statement, 293–296
    LM34-based controller, 296
    overview, 181
    reading thermocouples, 298–299
    thermistor-based controller considerations, 296–298
slave port functions, PORTE, 32–33
smell, 159
sockets
    EEPROMs, 115
    U10
        making sure working, 134–137

sockets (*Cont.*):
overview, 131–133
using RS485 communications, 137
U3, 115–117
U4, 113–115, 117–118
U5, 113–115, 118–120
U6
clock ICs, 122–124
DS1302 real-time clock, 123–124
LTC1298 12-Bit A-to-D converter, 124–126
overview, 120–122
U7, 126–129
U8, 126–129
U9
making sure working, 134–137
overview, 131–133
using RS485 communications, 137
software compiler, PIC 16F877A microcontroller unit, 7
software setup
loading, 15
notes from microEngineering Labs, 18–19
overview, 13–14
using in Windows environment, 15–18
solar data collector project
controlling fan, 310
microcontroller hardware, 304–305
overview, 182, 301–304
software
overview, 306–307
reading sensors, 309–310
setting internal clock, 307–309
updating LCD
overview, 310
problem, 314
sending data to computer, 311–314
solder joints, 316
soldering iron accessory, 332
solderless breadboards, 320
solid state relays (SSRs), 174, *175f*
speakers, 31–32
speedometers, 160, 332

SPI (serial peripheral interface) memory, 115, 117–118
SPST switches, 185, 293
SSRs (solid state relays), 174, *175f*
stabilized output, 184
startup, LCD, 322
steel tape, 161
switches, SPST, 185, 293
syntax errors, 17

T1CON control register
bit assignment, 100
bits used to control Timer1, 224, 226
functions of, 108
T1OSCEN control bit, 257
T1SYNC control bit, 256
T2CON control register, 102–103, 228–229
table mechanisms, 276–277
tachometer project
detection, 192–193
high speed considerations, 191–192
low rate considerations, 189–191
options, 187–189
overview, 179
seven-segment displays, 195–207
software, 194–195
tachometers, 178, 221–224, *225f*
taste, 159
telephone key tones on piezo speaker, 63
temperature sensors, 185
ten-pin connectors, 266, 271
thermistor-based controllers, 296–298
thermocouples, 298–299
thermometers, 160, 178–179
thermostats, 160
Timer0, 84, 104, 211, 214–224, 246–248
INTCON (interrupt control register), 89–90
LCD clock program using ON INTERRUPT, 90–93
OPTION_REG (option register), 88–89
overview, 85–88
using as counter, 105–107
Timer1, 84, 104
marble counter project

Timer1 (*Cont.*):
in asynchronous counter mode, 256–257
in counter mode, 248–255
INTCON interaction, 258
oscillators, 257
prescalars, 257
resetting register pair (TMR1H, TMR1L), 257
resetting using CCP trigger output, 257
metronome project, 211, 224–227
overview, 93–100
prescalars, 100
register pair (TMR1H, TMR1L), 257
running critical interrupt-driven tasks, 100–102
using as counter, 107–111
Timer2, 84
making sure working, 104
marble counter project, 258
metronome project, 212, 228–230
overview, 102–104
watchdog timer, 104
timers. *See also* Timer0; Timer1; Timer2
counters, 34
exercises, 112
overview, 33–34, 83–85
prescalars and postscalars, 34
watchdog, 34
TMR1H control register, 108
TMR1L control register, 108
touch, 159
touch panel project
building, 282–285
exercise, 291
hardware needs, 281
overview, 279–281
software, 286–291
software needs, 281–282
touch screen project, 179, 181
transducers. *See* sensors
TRIS registers, 325
TRISA register, 29–30
TRISB register, 30–31
TRISC register, 31
TRISD register, 32
TRISE register, 32

TTL level signals, 169, *170f*
two-axis software, 275

U10 socket
  making sure working, 134–137
  overview, 131–133
  using RS485 communications, 137
U3 socket, 113–117
U4 socket, 113–115, 117–118
U5 socket, 113–115, 118–120
U6 socket
  clock ICs, 122–124
  DS1302 real-time clock, 123–124
  LTC1298 12-Bit A-to-D converter, 124–126
  overview, 120–122
U7 socket, 126–129
U8 socket, 126–129
U9 socket
  making sure working, 134–137
  overview, 131–133
  using RS485 communications, 137
universal instruments
  basic temperature controlling device, 184–186
  notes, 186

universal instruments (*Cont.*):
  properties and capabilities of, 183–184
USB port programmer, 14

variables, 317
voltage, 28, 166–167
voltmeters, 160
volt/ohm meters (VOMs), 183–184
voucher, microEngineering Labs, 16

wall thermometers, 332
watchdog timers (WDTs), 34, 89, 215
  metronome project, 230–231
  Timer2, 104
Web site
  PICBASIC PRO programs, 345
  project, 177
  support, 349–350
Wi-Fi Internet service, 10
Windows environment, software use in, 15–18
wiring
  checking, 316
  diagrams
    counting into counter TMR1, *249f*

wiring (*Cont.*):
  IR-LED phototransistor pairs, *240f*
  programmable pulse counter, *201f*
  set of four seven-segment displays, *197f*
  seven-segment displays, *199–200f*
  single IR transmitter—phototransistor pairs, *284–285f*
  single-point programmable controllers, *295f*
  solar collector instrument, *305f*
  testing sensitivity of phototransistor response, *254f*
  testing TMR1 counter, *253f*
  touch panels, *282f*
  touch panel, *281f*
writable counting registers, 102
Write Data command, *156t*
writing to clock program, 308–309

XT setting, 324

Exclusive offers from microEngineering Labs, Inc.

Use the coupons below for big discounts on melabs development tools. Contact us for more information.

www.melabs.com

Phone: (719) 520-5323   Fax: (719) 520-1867   Box 60039, Colorado Springs, CO 80960

Valuable Coupon

Developer's Bundle including PICBASIC PRO™ Compiler, LAB-X1 Experimenter Board, melabs U2 Programmer, PIC® MCU, Cables, and AC Adapter.

All for only $499.95

To use this coupon, call 719-520-5323 or email support@melabs.com to order.

Discount excludes shipping and sales tax. Cannot be combined with other offers. Offer may be canceled or changed without notice. Coupon valid only if obtained from original publication. Void If sold, exchanged, or obtained by download. Cannot be applied to prior purchases. Not redeemable for cash.

Valuable Coupon

Valuable Coupon

melabs U2 Programmer with Accessories
plus
PICBASIC PRO™ Compiler

Both for only $349.95

To use this coupon, call 719-520-5323 or email support@melabs.com to order.

Discount excludes shipping and sales tax. Cannot be combined with other offers. Offer may be canceled or changed without notice. Coupon valid only if obtained from original publication. Void if sold, exchanged, or obtained by download. Cannot be applied to prior purchases. Not redeemable for cash.

Valuable Coupon

CPSIA information can be obtained
at www.ICGtesting.com
Printed in the USA
LVHW101539250620
658994LV00005B/274